# ROMANTIC BIOLOGY, 1890–1945

History and Philosophy of Biology

*Series Editors:* *Dawn M. Digrius*
*Rasmus Grønfeldt Winther*

# ROMANTIC BIOLOGY, 1890–1945

BY

Maurizio Esposito

LONDON AND NEW YORK

First published 2013 by Pickering & Chatto (Publishers) Limited

Published 2016 by Routledge
2 Park Square, Milton Park, Abingdon, Oxfordshire OX14 4RN
711 Third Avenue, New York, NY 10017, USA

First issued in paperback 2015

*Routledge is an imprint of the Taylor & Francis Group, an informa business*



BRITISH LIBRARY CATALOGUING IN PUBLICATION DATA

Esposito, Maurizio, author.
Romantic biology, 1890–1945. – (History and philosophy of biology)
1. Organisms. 2. Biology – Classification. 3. Biology – Philosophy.
I. Title II. Series
570.1'2-dc23

ISBN-13: 978-1-138-66228-5 (pbk)
ISBN-13: 978-1-8489-3420-7 (hbk)

Typeset by Pickering & Chatto (Publishers) Limited

# CONTENTS

# ACKNOWLEDGEMENTS

Several people, institutions and countries have directly or indirectly contributed to this work. First of all, I have to thank Gregory Radick for inspiring this book and fostering its achievement. I also thank all the staff of the Centre for History and Philosophy of Science at the University of Leeds, who provided a beautiful, friendly and productive environment throughout my stay. In particular, I thank Jonathan Hodge, Graeme Gooday, Adrian Wilson, Annie Jamieson, Steven French, Efram Sera-Shriar, Berris Charnley, Emanuele Archetti, Jonathan Topham and Chris Kenny. I am also grateful to all the staff of the Brotherton Library and Boyle Library at the University of Leeds.

I am deeply indebted to the Institute of Philosophical Investigation at the National Autonomous University of Mexico for its generous support. I have spent a wonderful and stimulating year there working on this monograph. In particular, I thank Carlos López Beltrán, Abigail Nieves Delgado and all the staff of the institute. I also thank all the participants of the Philosophy of Biology seminar at the Metropolitan Autonomous University, Cuajimalpa, Mexico, where a part of this work was presented and discussed.

Many other people gave me large or small advices, comments and inspiration along the way. In particular, I remember Joe Cain, Richard Delisle, Ricardo Noguera Solano, Thierry Hoquet, Juan Manuel Rodríguez Caso, Jan Sapp, Garland Allen, Jane Maienschein, Manfred Laubichler, Edna Maria Suárez Diaz, Christian Reiss, John Betty and Mark Ulett.

I am grateful to the British Society of the History of Science; the History of Science Society; the International Society for the History, Philosophy and Social Studies of Biology; Arizona State University; the University of California, San Diego; and the Société d'Histoire et d'Epistémologie des Sciences de la Vie for their generous support in different stages of my research. I thank the editor of this series and the three anonymous referees for their precious comments and invaluable critical advice. I thank the Department of Philosophy at the University of Santiago, Chile, where I have completed this monograph. A final special thanks go to my parents, Enza Varriale and Corrado Esposito.

## Copyright Permissions

I thank the following institutions and individuals for granting me permission to mention and/or publish archival sources: the Scripps Institution of Oceanography in San Diego, in particular the director of the Scripps Historical Archives in La Jolla, Peter Brueggeman, and the archivists Rebecca Smith and Carolyn Rainey; the staff of Bancroft Library at the University of California, Berkeley; the staff of the Woods Hole Historical Collection Archives, in particular the assistant director, Diane Rielinger, and the archivist Lindsey Fresta; the UCL Special Collections Library, in particular the archivist Mandy Wise; the Special Collections Department of the University of St Andrews Library; and the Zoological Station Anton Dohrn of Naples, in particular the archivist Massimiliano Maja.

## Credits

Permission to use unpublished material by D'Arcy W. Thompson granted by Moira Mackenzie, University of St Andrews, Special Collections Division, UK.

Permission to use unpublished sources by F. R. Lillie and E. E. Just granted by Diane Rielinger, Woods Hole Marine Zoological Station, USA.

Permission to use unpublished materials by W. E. Ritter granted by Susan E. Syder, Bancroft Library, University of California, Berkeley, USA; Linda C. Claassen, Special Collections, University of California, San Diego, USA; and Massimiliano Maja, Historical Archives, Zoological Station Anton Dohrn, Naples, Italy.

While every effort has been made to secure permission for copyrighted material, and to credit the material as required by copyright holders, the author will endeavour to correct any unintentional oversights.

# INTRODUCTION

If you want to describe life and gather its meaning,
To drive out its spirit be your beginning,
Then though fast your hand lie the parts one by one,
The spirit that linked them, alas is gone,
And 'Nature's Laboratory' is only a name,
That the chemist bestows on it to hide his own shame.[1]

J. W. von Goethe, *Faust*

Romanticism is an epoch. The Romantic is a state of mind not limited to one period. It found its fullest expression in the Romantic epoch, but it does not end with that age; the Romantic exists to the present day[2]

R. Safranski

At the end of the nineteenth century an oft-quoted sentence circulated among critics of Matthias Jakob Schleiden's cell theory and was usually attributed to the German botanist and mycologist Anton de Bary (1831–88): 'Die Pflanze bildet Zellen, nicht die Zelle bildet Pflanzen' ('The plant forms cells; the cell does not form plants').[3] Although in the beginning the aphorism was employed against the widespread practice (since the development of Schleiden's cell theory) of starting textbooks on botany with the study of cells rather than of whole plants, its use was quickly extended. In the hands of embryologists and physiologists, the aphorism acquired a polemic charge against the mechanistic interpretation of organic development and indicated a holistic way to conceive the process of morphogenesis. In Europe and the United States the sentence spread and was repeated as a refrain by many first-rank zoologists working in old or emerging institutions. One of the most convinced advocates of de Bary's idea was the influential nineteenth-century American biologist Charles Otis Whitman (1842–1910). Indeed, in 1893 Whitman published an article that inspired many young biologists who were critical of mechanistic and 'elementalist' interpretations of living phenomena. The article, titled 'The Inadequacy of the Cellular Theory of Development', began by criticizing Schleiden's cell theory.[4] Cells should not be seen as individual or elementary entities able to produce and form complex organs and tissues during morphogenesis. Rather, a higher guiding or formative principle had to be presupposed. For Whitman, in fact, the whole organization shaped and directed cellular proliferation and regulation during development: 'Development, no less than other vital phenomena, is a function of organization ... the formation of the embryo is not controlled by the form of cleavage. The plastic forces heed no cell-boundaries, but mould the germ-mass regardless of the way it is cut up into cells'.[5] Regenerative phenomena supported such an interpretation, because the reproduction of both *Hydras* and *Stentors* showed that 'the organism dominates cell-formation, using for the same purpose one, several, or many cells, massing its material and directing its movements, and shaping its organs, as if cells did not exist, or as if they existed only in complete subordination to its will'.[6] Whitman defined this view as the 'organism-standpoint', a stance he considered prominent among botanists but which he deemed extendable to the whole living kingdom.

In reinforcing what de Bary and other naturalists had stressed a few decades before, Whitman was emphasizing how reductionist approaches were misplaced when applied to living beings. Indeed, in the organic realm the whole had a priority over the constituent parts; the parts were explained, shaped and organized according the properties of the entire organism during its development. To Whitman, and to many of his predecessors and followers, this idea involved some fundamental descending corollaries: that biology had its independence from physical sciences; that life could not simply be reduced to physico-chemical substances; and that organisms could not be seen as mere mechanisms because of their irreducible interactions and the reciprocal relations between the parts and the whole. Yet – and precisely because of these relations – organisms had to be seen as teleological entities: if the whole was more than the parts, the parts function in terms of the whole. The parts were elements playing a specific role in an interactive system and working for the welfare of the whole. Teleology was the result of a dynamic kind of relationship between the whole and its components. Finally, the organism had to be seen in a constant and active relationship with the environment. The internal and external environment constituted and explained the organism ever since its earliest developmental phases. This conviction entailed the epistemic assumption – which as we will see was much diffused among organicist biologists – that fieldwork and laboratory studies were complementary; both were required in biological investigations precisely because an organism out of its environment was considered a comfortable abstraction. All these ideas supported each other and together constituted what many anti-mechanistic and anti-reductionist biologists called, during the first decades of the twentieth century, 'organismal or organicist biology' – a research programme that involved institutions, a network of renowned naturalists and a still-vibrant 'Romantic' bio-philosophical tradition.

This book is about such a research programme and the people, institutions and traditions furthering it. Yet, and more generally, the book is also about the way the organicist tradition, in its diverse forms and manifestations, influenced new scientific problems, experimental practices and controversial results during the first decades of the twentieth century. In drawing an intellectual map of influences, connections and ideas, the book also charts the different ways through which an old eighteenth- and nineteenth-century organicist tradition – a Kantian and Romantic tradition – was filtered and re-flourished in the twentieth century. We will see how important Romantic naturalists and philosophers were for the formulation of an organismic programme; how influential their ideas and speculation on the nature of life were for some twentieth-century biologists; and how relevant their views were on how organisms had to be perceived and studied. In other words, we will explore the ramifications, interpretations and multiple waves of diffusion of such a Romantic, holistic bio-philosophy in England and the United States, without overlooking its circulation in other countries.

Thanks to the efforts of many historians and philosophers in the last five decades, we have a fairly good knowledge about eighteenth- and nineteenth-century Romantic sciences. From the pioneering work of Alexander Gode-von Aesch, *Natural Science in German Romanticism* (1941), and Stephen J. Gould, *Ontogeny and Philogeny* (1977), many more recent contributions followed. From the book edited by A. Cunningham and N. Jardine, *Romanticism and the Sciences* (1990), to other collections of essays and books, individual articles and companions, we are aware that German Romanticism and further forms of Romanticism in other countries were deeply preoccupied with scientific issues.[7] Eighteenth- and nineteenth-century Romantic naturalists provided an important and enduring contribution to the sciences, from physics to biology, from geology to natural history. As John Reddick emphasizes, Romantic naturalists and *Naturphilosophen* 'did make a very real contribution to the development of the sciences ... in particular, its dare-to-speculate mentality greatly furthered the sciences in their quantum leap from the physics derived fixed-mechanism model of the world in the eighteenth century, to the transformational, evolutionary model so characteristic of the nineteenth'.[8] A few years later Stefano Bossi and Maurizio Poggi followed: in the introduction of their edited book *Romanticism in Science* (1994), they explain: 'we want to make it very clear that the Romantic age was a great scientifical age. It is very important to underline this and to avoid all misunderstandings in front of the widespread nostalgic attitude of those who look back to the first decades of the last century, in search of some mysterious, oracle-like, prophetic, or even catastrophical elements'.[9] Frederick Beiser, in a convincing attempt to evaluate *Naturphilosophie* in its appropriate historical context, also recounts that the *Naturphilosophen* were not the over-speculative, anti-empirical and irrational 'cranks' that have often been portrayed. *Naturphilosophie* was considered by many of his practitioners a 'serious' science, with grounded methods, results and theories often based on experiments and observations.[10]

Thus this book owes a great debt to the immense literature about the Romantic sciences. Thanks to all these narratives, I could compare, analyse and assess continuities and differences between nineteenth-century Romantic organicisms and twentieth-century organismal biologies. I could understand and reconstruct institutional connections and intellectual influences that nineteenth-century naturalists exerted on some of their close and distant followers. I could indeed evaluate and track down how some Romantic discourses about life and the way it should be investigated were progressively translated, modified and reframed, then decontextualized and recontextualized in other countries and intellectual worlds. In other words, I was able to explore how old ideas and notions were employed and deployed for inventing new coherent and influential organicist discourses – in particular, how the Romantic metaphor of the universe as a dynamic and creative organism was resurrected and re-employed in diverse, more focused, discussions; how the biogenetic law that had characterized many speculations of Romantic naturalists was reframed and re-adapted to new findings and theories in the twentieth-century life sciences; and finally, how a new form of typological approach continued to be defended and advanced for the understanding of evolution and development.

Different scholars have already outlined the importance of the organicist traditions in the twentieth century from different perspectives. According to Ludwig von Bertalanffy, Morton Beckner, Ernst Mayr and Donna Haraway, most of these ideas and approaches characterizing organismal biology constituted a new twentieth-century paradigm in life sciences. Von Bertalanffy felt himself to be an architect of a new organicist paradigm. As he wrote back in the 1920s, 'We believe that the attempts to find a foundation for theoretical biology point at a fundamental change in the world picture. This view, considered as a method of investigation, we shall call "organismic biology"'.[11] Beckner maintained that the distinction between mystical or vitalistic views and organismic approaches 'was drawn clearly only in the twentieth century. Organismic biology may be described as an attempt to achieve the aims of the murky organismic-vitalistic tradition, without appeal to vital entities'.[12] Mayr followed by noting that organicism was an effective attempt to overcome the dichotomy of mechanism versus vitalism: 'When organicism, the new holistic philosophy of biology, developed in the twentieth century, particularly under the influence of Darwinism, the previous polarity became invalid'.[13] According to Haraway, such a distinction could be explained in Kuhnian terms: the new organicist paradigm was the result of a period of crisis in developmental biology. She argued that the first half of the twentieth century was a time when 'the age-old dichotomy between mechanism and vitalism was reworked and a fruitful synthetic organicism emerged, with far-reaching implications for experimental programmes and for our understanding of the structure of the organisms'.[14] In particular, she pointed to Driesch and his well-known experiments on sea urchin eggs as the event that highlighted an 'anomaly' within the nineteenth-century mechanistic paradigm in biology. To Haraway, Driesch had shown that the extraordinary developmental plasticity of sea urchin embryos apparently defeated any mechanist or reductionist explanation. Thus a new paradigm was required: a non-vitalist organicism.

However, drawing on Philip Ritterbush's *Art of Organic Form* (1968), Haraway recognized the importance of pre-holistic and pre-organicist traditions stemming from Kant's and Goethe's bio-philosophical thought. But because she genuinely believed in the 'new paradigm' thesis, she was unable to see any relevant historical tie between late eighteenth- and nineteenth-century holism and twentieth-century organicism. Such a historical link was clearly seen by Scott Gilbert and Sahotra Sarkar.[15] For them, organicism had been, ever since Kant, the philosophy of the embryologists and developmental biologists. Nevertheless, they were far more interested in justifying organicist epistemology for contemporary biology than in examining genealogical connections. Fifteenth years before, Ernst Cassirer proposed a fascinating link between Kant's bio-philosophy, Goethe's holistic science and some of the twentieth-century developments and approaches to life sciences. In particular, in mentioning the German botanist Emil Ungerer, who had dedicated a book to Kant's notion of teleology,[16] Cassirer noted that 'the latest phase of biology more than any other prior to it has brought into currency the fundamental view which Kant had advocated. The modern movement called holism or organicism agrees in its general orientation with the methodological tendency of Ungerer'.[17] Such a fundamental view, Cassirer concluded, was a view that J. S. Haldane from England, von Bertalanffy from Austria and Ungerer, Julius Schaxel and Jakob von Uexküll from Germany were diffusing and defending in the scientific community.

More recently, A. Harrington has even more clearly established the connection between romantic holism and the organicist view in the life sciences. According to her, some expressions of German holism in the twentieth century heavily drew on Kant's aesthetic view of the organic world and Goethe's Romantic conception of 'wholeness'.[18] Analysing the lives and careers of four leading organicists and controversial figures, the neurologist Constantin von Monakov (1853–1930), the behaviourist von Uexküll (1864–1944), the psychiatrist K. Goldstein (1878–1965) and the psychologist M. Wertheimer (1880–1943), she illustrates how much twentieth-century holism owes to the Kantian and Romantic world. In addition, she makes another profound and fascinating connection: some holistic discourses and approaches in Germany backed and supported fascist and Nazi ideologies. As she emphatically recounts, the Romantic idea of wholeness was deliberately translated into practical policies by some National

Socialist enthusiasts. In fact, one Nazi biologist, H. J. Fuerborn, intended to use the holistic doctrine for didactic purposes. As Harrington reports: 'the core of all biological education in the Nazi schools could be found in three basic principles: the doctrine of biological wholeness (the whole as greater than the sum of its parts), the theory of biological development (the dynamic creation of organismic wholes), and the teachings of heredity (the transmission of the qualities of the whole across generations)'.[19] We will see that almost all of the figures I consider in this book supported and proposed very similar ideas, although with very different purposes and results.

Thus, in line with Cassirer and Harrington, and contrary to von Bertalanffy, Beckner, Mayr and Haraway, I will argue that organismic or organismal biologies have a very long pedigree, a history that can be linked to Aristotle's zoology, Kant's bio-philosophy (as largely articulated in his *Kritik der Urteilskratf* published in 1790) and that of Kant's Romantic followers.[20] However, I will tell an original story different to that of Cassirer and Harrington. I will track the diffusion and uses of this particular philosophy in the United Kingdom and the United States from the nineteenth century and the different impacts it had on specific institutions and a range of individuals, as well as the international ramifications. Through the analysis of textbooks, books, articles and archival resources (letters and unpublished papers), we will view the unexpected emergence of an international community of scientists – British and American, of course, but also French, German and Italian – all committed to spreading and supporting organismic and holistic views in biology and society. An entire intellectual continent will emerge, a partially forgotten world buried in dusty archives and dark library stacks. We will discover that a sophisticated and original philosophy of biology was formulated, discussed and internationally accepted during the very first decades of the twentieth century. This philosophy, we will see, was not rooted in the work of well-known heroes such as Darwin, Spencer or Mendel; instead it was rooted in and inspired by the studies of German and French embryologists, morphologists, anatomists and physiologists. These were people such as von Baer, Goethe, Georges Cuvier, Johannes Müller and Rudolf Leuckart, as well as influential philosophers such as Henri Bergson and Alfred North Whitehead. In other words, we will explore an almost pristine (or at least very partially known) territory in the history of science.

We will become aware of the impact of such a philosophy on some controversial views that many of these scientists held about contemporary bioscience. Many of those I include in my narrative proposed a coherent synthesis that included heredity, development and evolution. Differently from many advocates belonging to the tradition of modern evolutionary synthesis – who often regarded gene selection and adaptation as the central elements explaining speciation – their synoptic alternative saw organic development as the chief phenomenon to investigate in order to understand evolutionary novelties. Back in 1920 a British evolutionary embryologist, Walter Garstand (1868–1949), synthesized the spirit of this alternative with a lucid proposition: 'Ontogeny does not recapitulate phylogeny: it creates it'.[21] As we will see, Garstand's conviction was widely accepted by many biologists active during the first four decades of the twentieth century. Interestingly enough, this is a view that continues to be accepted by some developmental biologists. As Robert Lickliter and Thomas Berry commented in 1990, 'phylogeny is not a sufficient causal or mechanical explanation for ontogeny. Rather, phylogeny is best characterised as an historical description of progressive variants in ontogenesis'.[22] In the Conclusion of this book it will be shown that theorists of the 'evo-devo' – a new discipline born in the 1980s – share many elements of the early twentieth-century Romantic tradition of life science.

Finally, we will notice that many figures sharing organismic or holistic approaches to biology and society, both in the United Kingdom and the United States, were very far from supporting fascist, totalitarian or rightist ideologies or convictions. On the contrary, most of them were fully engaged in fostering progressivist or leftist causes; they genuinely believed that the 'organism as a whole' represented a powerful metaphor for a well-ordered, fair, liberal and democratic society. Richard Lewontin and Richard Levins have argued that reductionist and mechanistic approaches are the hallmark of the bourgeois ideology.[23] According to such a 'Cartesian' ideology, individuals or systems are better explained through a reduction to their composing parts. Conversely, a science based on dialectical principles – principles inspired by Marxist dialectical materialism and organicism – can better grasp the complexity of natural processes. Although Levins and Lewontin's distinction can appear convincing from a philosophical viewpoint, it is flawed from a historical perspective. Indeed, holistic and organicist views about biology and society were rather associated with diverse ideologies. When we look seriously at the history of biology, we would easily find Marxists upholding Cartesian and reductionist approaches in science (Loeb, Müller or Hogben for instance) and, on the other hand, reactionaries spreading holistic models of causation (von Uexküll or von Monakov). In other words, history teaches us that there was not a necessary relationship between particular political creeds, social ideologies and scientific visions. The complex historical connections between leftist or rightist doctrines and scientific theories are contingent. They are established in processes of contextual negotiation and did not follow from an intrinsic power of ideas.

Of course, this bio-philosophical tradition did not exemplify a paradigm, a static research programme, or a stable, unified and coherent set of beliefs that all adherents to such a movement uncritically shared. Instead, it was characterized by a flexible and open tradition in which diverse assumptions and problems were negotiated and interpreted in different ways. The way I use labels such

'Romantic', 'organicist', 'holistic' or 'anti-mechanistic' is regulative and purely descriptive. In 1924 Lovejoy had warned that the word 'Romantic' 'has come to mean so many things that, by itself, it means nothing'.[24] Almost a century later, we are not in a better situation. The word 'Romantic' has probably become even vaguer than in Lovejoy's time. However, the famous historian of ideas proposed a positive solution. Instead of formulating a new and controversial definition – or simply avoiding the word altogether – Lovejoy argued that we should use the notion of 'Romanticism' in the plural. We have different Romanticisms, where each 'was made up of various unit-ideas linked together, for the most part, not by any indissoluble bonds of logical necessity, but by a logical associative process, greatly facilitated and partly caused, in the case of the Romanticisms which grew up after the appellation "Romantic" was invented, by the congenital and acquired ambiguities of the word'.[25] If we pretend to find an 'essence' of the Romantic tradition, we would probably end up with a caricature; but if we accept a certain semantic vagueness, we can establish interesting connections between different research traditions, epochs and institutions. In other words, Romanticism does not dwell in some ghostlike intellectual roots; it dwells in the interstices of a changing network. In the spirit of Lovejoy's proposal, this book explores a cross-section of this network.

However, Isaiah Berlin, in reminding us that historical labels are problematic, mentions the French poet Paul Valery, who said, '*One cannot get drunk, one cannot quench one's* thirst, with labels on bottles'.[26] While it is true that you cannot get drunk with labels on bottles, this does not imply that labels are useless. Labels help you to know what you are drinking. They describe and indicate what kind of substance you are about to sip – an old or new wine, a grape juice or a very special kind of cognac. Labels give you useful information even if you merely want to get drunk or quench your thirst. Similarly, historical labels help you to understand what is going on in a given epoch or in a particular community. To use another analogy, a group of trees, a blue lake or a yellow straight line drafted on a map may orient the lost tourist onto the right track; historical labels similarly orient the reader in the vast, contradictory and complex network of ideas and traditions. Labels are like orienting symbols. Goethe used to say that historiography of science is like a 'labyrinthine garden'.[27] Events, people and the genealogy of particular ideas can be confused, intricate, circular. In my view, the role of the historian is not to provide a gardener's map having a scale of meter to the meter, but to offer instruments and some instructions for the reader's journey into the labyrinth of history. Thus my historical reconstruction must be seen like a map indicating a partially explored territory; it does not have the ambition of recounting what 'really happened' outside any possible perspective. As a consequence, when I use the label 'Romantic' I do not intend to reify a historical movement, i.e. to say that there were necessary conditions that included some members and excluded others. My aim is to portray an intellectual tendency, inclination, propensity, idiom, a particular 'will' or 'bias' to solve problems in a certain way and with a peculiar 'custom' of thinking. I previously tried to capture this custom through the characterization of Whitman at the start of this Introduction.

This book is the result of my growing awareness that many historical narratives about twentieth-century biosciences omitted many important elements. We know many things about some of the heroes of modern synthesis. We know a good deal about Darwin and the neo-Darwinian tradition. Historians have made great progress in reconstructing the intellectual and institutional landscape behind reductionist and mechanicist traditions in twentieth-century biology. But we know much less about the organicist tradition - about, for instance, Rudolf Leuckart, William Ritter, Charles Child and John Haldane. We also know very little about some alternative syntheses that less well-known figures forged during the heyday of the evolutionary synthesis. Although historians have begun to chart some of these alternatives – in extending the intellectual, institutional and geographical frameworks of the old narratives – we are far for having an adequate understanding of what biology was between the 1900s and 1940s, even in those countries where focused studies are widely available, such as the United States and the United Kingdom.[28] However, the aim of this book is not to merely fill up a small place on the bookshelf of the history of biology. I undertook my reconstruction, I gave voice to some overlooked figures, and I unearthed some forgotten institutions because I am deeply convinced that we – as historians, as philosophers, as biologists – can still learn important things from them. This is especially true because in the very last few decades organicist ideas have been re-emerging and organismal biology is becoming an established research programme in many departments around the world. As the American biologists K. Schwenk and his colleagues assured in 2009:

> Organismal biology is currently experiencing a renaissance. Some have considered it old fashioned, and not an integral part of 'modern biology'. It is increasingly clear, however, that organisms are the key to organizing and understanding the information in other areas of biology ... There is increasing recognition that to understand life and basic processes, biological information must be understood in the context of the integrated living organism, and not as a collection of systems operating independent of the organism. Indeed, organisms are the central biological unit that integrates and responds to internal and external information.[29]

Critiques of gene-centric views of development and evolution are now common. Notions such as hierarchy, field, system and downward causation are once again on the stage. Evo-devo has given new currency to the old idea that developmental mechanisms can give important insights into the understanding of speciation. Indeed, my bet is that developmental biologists, philosophers and evolutionary biologists – especially those fascinated by the new issues linked to the extended modern synthesis – can find very useful ideas, informed critiques and new perspectives in some of these old organicist proposals.

Any historical narrative is characterized by inclusions and exclusions. Mine is not an exception. I focused on a specific community, institutions and countries.

But the rationale behind my selection was not predetermined from the beginning; the structure of my account was not, so to speak, fixed and then filled with biased evidence and partisan anecdotes. Both the structure and stories grew and developed gradually out of my general interests in developmental and evolutionary biology, my archival investigations, my understanding of the nineteenth- and twentieth-century life science, my reading and rereading of the primary sources, and my critical assessment of the secondary sources. Essentially, the specific selection of figures constituting my community of scientists is empirically based. It is the result of the correspondence-network I found in the archives and my textual analysis of published works. As a methodological maxim, I utilized five levels of evidence: 1) correspondence, 2) friendship, 3) institutional relations, 4) reciprocal admiration demonstrated in both conferences and publications, and 5) only as a final element, ideas and interests in common. At the end of my general assessment of the evidence acquired, I selected eight figures who all had special ties (intellectual, institutional, personal, etc.), making them a very interesting unit of historical investigation – a unit that, nevertheless, displayed an interesting multiplicity: i.e. eight different ways of doing and understanding organismal biology.

Yet I am completely aware that I left out important figures and institutions. I surely overlooked important books, articles and other archives. This book is very far from being a comprehensive history of organicist philosophy. I am very conscious that this study is, from a general point of view, incomplete. I see it as an open manuscript left to be improved and advanced by other academics. I hope that other historians, starting with this book, can produce conceptual maps more exhaustive than mine, maps drawn from even more complex narratives, especially focused on the German, French and Russian worlds.

The chapter organization is as follows. The first chapter explores Kant's bio-philosophy, its multiple interpretations and its progressive 'Romanticization' in Germany and elsewhere. In the second chapter I chart the diffusion and transformation of such a Romanticized bio-philosophy in the late nineteenth and early twentieth centuries. In particular, I show what kinds of relations – historical, institutional and philosophical – Kant has with nineteenth-century Romantic naturalists, and then twentieth-century biologists. In the third and fourth chapters I illustrate the influence of post-Kantian and Romantic bio-philosophy in England during the twentieth century as expressed by two generations of relevant British biologists: J. S. Haldane, D'Arcy W. Thompson, E. S. Russell and J. H. Woodger. Finally, in the fifth and sixth chapters I assess the impact of such a tradition in the United States. In particular, I analyse the work of W. E. Ritter, C. M. Child, F. R. Lillie and Lillie's student, E. E. Just, and demonstrate how it affected the agendas of whole institutions in which they worked or which they directed. Throughout each chapter, I explore how this 'Romantic' tradition, transformed in organismic philosophies during the twentieth century, was associated with current political and ideological doctrines. In the Conclusion I try to answer one important question: why did organismal biology almost disappear? I also provide a general overview of what happened after World War II, and outline the progressive re-emergence of organismal ideas after the 1970s. We will see the important differences, translations and continuities between the old conception of organismal biology and our contemporary – more specific and circumscribed – version.

# 1 OLD AND NEW ORGANICISMS

> [T]he assumption that the orderly course of a process can be represented by an analysis of it into temporal and spatial processes must be dropped. It is thus the concept of wholeness which must be introduced as well as into the field of physics as into that of biology in order to enable us to understand and formulate the laws of nature.[1]
>
> Max Planck

## London, 1931

In late June 1931 W. E. Ritter, professor of zoology at the University of California, Berkeley and former director of the Scripps Institution of Biological Research, arrived in London in order to chair the third session of the Second International Congress of the History of Science.[2] The London meeting is mainly remembered as a gathering of radical scientists discussing how science could shape capitalism and how scientific rationality could revolutionize Western societies.[3] However, the congress also included a small group of scientists with less ambitious goals: they aimed to rethink biological sciences on new theoretical bases. The title of Ritter's session was 'Historical and Contemporary Relationships of Physical and Biological Science', and participants included the physiologist J. S. Haldane, the zoologist E. S. Russell, the biochemist J. Needham, the engineer L. L. Whyte, the botanist B. Becking, the zoologist L. Hogben, the theoretical biologist J. H. Woodger, the zoologist D'Arcy W. Thompson and the physicist A. Yoffe.[4] In general, as Ritter remarked in his critical summary of the papers, the discussion touched on diverse issues, including: the supposed difference between inorganic and organic natures; the possibility of a scientific synthesis of the whole of natural knowledge; the importance of Aristotle and his tradition for renewed life science; the importance of observation and experiment in the biosciences;[5] and the new emerging physics exemplified by quantum theory, which, as Whyte put it, would transform 'certain aspects of the traditional antithesis between physical and biological theory'.[6] Different generations of scholars with very diverse backgrounds, convictions and views were discussing some of the most interesting scientific issues of the time. However, one of the central concerns coming up in almost all of these contributions was the allegedly peculiar status of life sciences, as contrasted with the theories and practices of other disciplines.

The first speaker in the session that June day, John Scott Haldane, was particularly concerned about the peculiarity of life sciences and its methods of inquiry. Indeed Haldane, a Scottish aristocrat who devoted all his life to physiology, argued that biology had to be considered an independent and irreducible science because, as he explained, 'each example of life ... can only be interpreted as the manifestation of a persistent and indivisible unity, recognized quite naturally and in common language as the life of the organism and the stock to which it belongs, and showing itself in endless coordinated details of form, environment, and the activity which express it'.[7] Haldane was in good company; Ritter himself, and more junior participants such as Russell and Woodger, fully agreed – both in this session and elsewhere – that biosciences rested on a different theoretical basis than disciplines like physics and chemistry. In fact, Russell was even more extreme than Haldane in expressing his uneasiness with physical methods as applied to investigations into life and, in particular, to his own discipline: animal behaviour. In a brief summary of his contribution, he contended that the influences of physics on biology had been quite negative in preventing 'the development of a real science of animal behaviour, and hence animal ecology'.[8] Woodger readily agreed; the modern biologist should start thinking more in terms of whole systems (as the philosopher Ludwig von Bertalanffy was teaching) than in terms of elementary 'stuff': 'modern biology requires a complex system of entities standing in abstractly specifiable relations to one another. Perhaps the notion of "protoplasm" or "living matter" will also go the way of "hereditary substance" when we learn to think of cells more in terms of systems, and less in terms of stuff'.[9]

Of course, not everyone agreed. For example, the young Cambridge biochemist J. Needham objected to Haldane in that between the physicist and biologist, there was a new, intermediate figure who should be regarded with interest by all scholars of biosciences: the crystallographer. As Needham asserted: 'The crystallographer deals with a form of organisation which is quite different from, and very much simpler than, the form of organisation which the biologist studies, but it is nevertheless organisation, a kind of rigid drill to which the aimless perambulations of particles in liquid of gaseous phases are subjected'.[10] If crystallography dealt with problems neither reducible to physics nor extensible to biology, Haldane's argument – according to which life sciences reflected a specific worldview – would imply that crystallography embodied a further irreducible worldview. As Needham ironically added: 'Crystal phenomena can only be studied in crystals, but I should not myself be in favour of adding a crystalline view of the universe to the biological view and the physical view, although I can see no logical reason, on Professor Haldane's side, for refusing this status to crystallography'.[11] In short, although different disciplines explained different aspects of the world, this did not imply that each discipline would be irreducible, in principle at least.

Others too disagreed with the position of Haldane and his close advocates; the Russian physicist Yoffe stated that there were already a wide array of biological phenomena studied through physical methods,[12] and the zoologist Hogben praised the virtues of mechanism and materialism as applied to modern biology as 'instruments for arriving at predictable conclusions about how organisms behave'.[13] Be that as it may, the end of the discussion brought a veritable conclusion on which almost everybody agreed: as Planck himself had stated in a short letter addressed to *Nature* in the same year, in light of quantum physics, the whole inorganic realm needed to be seen through a new perspective.[14] Classical physics, as exemplified by the traditional worldview and methods of Descartes, Galileo and Newton, was inadequate to explain the atomic world as well as insufficient to shed any light on the organic realm. Indeed, the new physics was based on systemic thinking and drew heavily on biological concepts.[15] Accordingly, even though life phenomena did not betray physico-chemical laws, no organism could be simply reduced to a mere physical mechanism. In short, biology had to be seen as the science studying complex systems rather than inert matter, activities rather than static structures, interactions rather than isolated entities, temporal processes rather than timeless beings, dynamic wholes rather than unconnected parts. Physics could not explain life because the living organism was not the mere sum of its parts.

This was not a new position. Haldane himself, during a presidential address at the physiological section of the British Association delivered in Dublin in 1908 (many years before the London congress), had emphatically and famously concluded his talk by claiming, 'That a meeting-point between biology and physical science may at some time be found, there is no reason for doubting. By we may confidently predict that if that meeting-point is found, and one of the two sciences is swallowed up, that one will not be biology'.[16] However, in the 1930s, when all of these debates were ongoing, the idea that biological science should have a special status (either ontological or epistemic) was losing credibility among the international community of biologists. The diplomatic conclusion according to which the inorganic and organic world had to be seen through the lens of 'wholeness' remained a very precarious and vague notion – a view easily forgotten by younger generations. Despite Haldane's authority, the wind blew towards reductionist approaches to biology. After the 1930s, the triumph of theories and doctrines perceived as reductionist (Morgan's chromosome theory of heredity, molecular biology and population genetics) challenged powerfully the purposes and prospects of 'systemic' thinkers. Yet before the London congress, 'system' thinking had a very positive and widespread reputation, a reputation supported by a significant tradition in the life sciences. Ritter called it 'organismal biology'. Others called it 'organicism'. But all knew that this was not a novelty. All knew that important naturalists and philosophers back in the eighteenth and nineteenth centuries had already expressed similar ideas.[17]

Needham had admitted that the concept of 'organicism' must be very old, although we ignore 'the genius of its original minting'.[18] Ritter rooted his 'organismal conception' in Aristotle, Kant, Cuvier and Goethe (see Chapter 6).

Haldane linked organicism with Kant, Müller and Scottish idealism. D'Arcy Thompson mentioned Goethe, Cuvier and Müller (see Chapter 3). Russell had no doubts that both Aristotle and Kant had to be credited with the idea, even though Kant was the philosopher who had given the most recent and complete formulation: 'The conception of the organism as essentially a unified whole is of course a very old one. It was clearly stated by Aristotle, and Kant has formulated it at great length in the *Critique of Judgment*'.[19] Indeed, most were aware that Kant had proposed, in his third critique, a sophisticated argument for overcoming the classic dichotomies such as preformation versus epigenesis and vitalism versus mechanism. Those who were historically informed also knew that some nineteenth-century Romantic naturalists, in following Kant's original proposal, invented a third way between mechanistic philosophies and vitalist speculations. However, before we can establish a convincing connection between twentieth-century organicism and eighteenth- and nineteenth-century holistic views – and in order to assess the importance of this tradition, the routes through which it travelled, the institutions and individuals it affected, and the way it changed over three centuries – we need to devote a few words to its original sources. We need to know what Kant himself believed and how his philosophy was reinterpreted by Romantic naturalists and thinkers.

## The Kantian Origins

It is well known that Kant himself speculated widely on the nature of organic beings in his 1790 work *Kritik der Urteilskratf*.[20] To Kant, the organism was a particular entity which required the presupposition of its purposive whole in order to understand the organization of its constituent elements.[21] Indeed, unlike minerals and other inorganic substances, organisms manifested a plan, a design, a complex organization that connected, in one integrated individual, parts and whole. Kant dubbed organisms 'natural products', where 'every part not only exists by means of other parts, but is thought as existing for the sake of the others and the whole, that is an (organic) instrument'.[22] However, organic unity was not enough for defining a 'natural product'. A living organism, unlike inorganic entities, clearly shows activities of self-generation and exhibits powers of self-organization. In approaching a general definition of the organism, Kant added that 'An organized being is then not a mere machine, for that has merely moving power, but it possesses in itself formative power of a self-propagating kind which is communicated to its materials though they have it not of themselves; it organizes them, in fact, and this cannot be explained by the mere mechanical faculty of motion'.[23] Therefore organic unity and the power of self-organization characterize an organism, an entity which can only properly be understood teleologically and not mechanically. However, Kant made things more complicated, because even though we need the notion of purpose to understand how organisms work, this did not imply that organisms were, ontologically, purposive entities. In order to clarify this last point, Kant distinguished between two different kinds of explanation: mechanical explanation (typical in physical sciences) and causal explanation (typical of natural sciences such as biology).

To him, mechanical explanations required a certain and definite chronological order between causes and effects: for example, billiard ball 1 causes the movement of billiard ball 2, so that ball 2 moves as an effect of the movement of ball 1. The cause–effect relation between the two balls is a linear and chronological one; we clearly know and distinguish which is cause and which effect. In the causal explanation, this is impossible. As Kant argued, in an entity with a natural purpose (*Naturzwecke*), the relations between causes and effects are never linear and chronologically determined for the simple reason that 'An organized product of nature is one in which every part is reciprocally purpose and means',[24] therefore, reciprocal cause and effect. Kant, however, added that in the difference between mechanical and causal explanation lay something else: different kinds of explanations required different types of judgements – *determinant* and *reflective*. Now, to Kant, the notion of *determinant* judgement denoted, in general, an intellectual faculty which connects the particular empirical data with universal laws given by the *a priori* category of the intellect. Thus, whereas 'determinant judgement' could be employed in a mechanical explanation because, as he had argued in his first critique, the objective knowledge about particular objects is *subsumed* under the *a priori* categories of the intellect (the linear and chronological relation between cause and effect, for example), no such thing was possible when that particular object was an organism. In that case, insofar as the particular object of knowledge cannot be *subsumed* under the universal *a priori* categories, the intellect requires *reflectierende Urteilskraft* (reflective judgement). In other words, in the case of reflective judgement, the intellect needs to find a proper regulative principle to make sense of the experiences made on a particular object or process when such an object or process, like an organism, implies reciprocity between causes and effects, parts and wholes. Yet that regulative principle is neither in the empirical data nor an intellectual faculty; it is a heuristic principle and, as such, is purely subjective. Therefore, this heuristic principle was seen by Kant as subjective judgement about natural things, a judgement which attributes particular purposes and ends (hence functions) to natural entities. In short, to Kant, we need teleology to make sense of life, even though organisms can be the mere outcome of mechanical laws. As he clearly argues, explaining the constitutive role of teleological judgement in the life sciences:

> According to the constitution of the human Understanding, no other than designedly working causes can be assumed for the possibility of organized beings in nature; and the mere mechanism of nature cannot be adequate to the explanation of these

> its products ... This is only a maxim of the reflective, not of the determinant judgment; consequently only subjectively valid for us, not objectively for the possibility of things themselves of this kind.[25]

It is because our intellect is limited by its universal *a priori* categories, Kant argued, that we need a subjective principle obviating our intellectual narrowness when we observe organisms. As R. J. Richards aptly puts it: 'according to the Kantian system, we apply categories like causality and substance determinatively to create, as it were, the phenomenal realm of mechanistically interacting natural objects. But in considering biological creatures, we must initially analyze the anatomical parts in reflective search of that organizing idea that might illuminate their relationship'.[26] However, to Kant, this did not mean that mechanical explanations had no place in life sciences; on the contrary, he thought that the naturalist should push as far as possible mechanistic understandings of living beings: 'The greatest possible effort, even audacity, in the attempt to explain them mechanically is not only permitted, but we are invited to it by Reason; notwithstanding that we know from the subjective grounds of the particular species and limitations of our Understanding that we can never attain thereto'.[27]

But why define the organism in such a way? Why, in other words, did Kant conceive of organisms as entities in which causes and effects are reciprocally entwined? Indeed, as we have seen, from that definition he derived the fact that mechanical explanation (and therefore determinant judgement) was not applicable to organic entities. Yet the core question is: from where did Kant get such a definition? This is a question less obvious than many Kantian interpreters would admit, but it is a strategic one for my purposes here. Indeed, the way that Kant defined a being with natural ends – an organism – reveals which organic phenomena he deemed central and what characteristics had to be considered *conditio sine qua non* so that an organism *may be* catalogued as a *real* one (rather than a complex machine, for example). Therefore, if we want to know how and where Kant developed his definition of organism, we need to know the essential elements, abilities and phenomena he thought typified what we recognize as a living being. First, to him, organisms were dynamic things insofar as they showed properties that no inert thing, however complex, could show. There were three general properties: organisms reproduce, develop and regenerate, whereas machines do not. Kant believed that these three properties justified the gap between the use of a mechanical and a causal explanation; a careful reading of the sections dedicated to the philosophy of biology in his third critique is the best evidence.[28]

To put it another way, the problem was not to make sense of the organisms and their constitutions once their process of development was terminated; the problem was to understand the laws ruling development itself. In principle, we could understand the organism mechanically, studying its anatomy, measuring its growth, observing its functions and so on; however, we cannot explain, from a mechanical viewpoint, how the organism's body forms itself. The mechanical explanation – and, therefore, determinate judgement – cannot be applied to organisms because organisms reproduce, develop and regenerate, all dynamic processes requiring reflective judgement.[29] However, Kant probably did not observe these phenomena by himself, since he did not perform experiments; in fact, for this he was indebted to two of the greatest naturalists of the eighteenth century, C. F. Wolff (1735–94) and J. F. Blumenbach (1752–1840).[30]

Wolff was a German embryologist who graduated from the University of Halle in 1759. In the same year he published a very influential work based on his dissertation: *Theoria generationis*. The book represented an epigenesist manifesto against the preformationist views about development. Beginning with plant growth and ending with animal embryology (in particular the formation of the intestines), Wolff postulates the existence of a force that he dubbed *vis essentialis*, a vital force explaining the creation and emergence (epigenesis) of new structures during organic development.[31] Recently, P. Huneman has convincingly shown that Wolff's *vis essentialis* – as an epistemic strategy for descriptive embryology – was behind Kant's reflexive judgement.[32] To Wolff, the postulation of a specific force behind development guaranteed the continuity of the embryological process. *Theoria generationis* offered Kant a strategic sample for his philosophical solution. Organisms were neither simply preformed in the egg nor were capricious creations; they were self-organized entities. During their development, they followed an ordered path that seemed to conform to a final end, *as if* a purposive force informed and directed ontogeny.

Blumenbach, in contrast, was primarily a physiologist and anatomist (though he is also celebrated as anthropologist); he, like many naturalists with biological interests at that time, was involved in the debate between followers of the preformationist doctrine and advocates of the epigenesist hypothesis. A previously committed preformationist, Blumenbach turned to epigenesis mainly because he felt that preformation could not explain the fact that mixed parents begot blended offspring. However, what Blumenbach had in mind while writing his influential 1781 treatise on epigenesist, *Uber den Bildungstrieb und das Zeugungsgeschafte* (The Formative Drive and its Relation to the Business of Procreation), were the staggering powers of a tiny creature that Linnaeus named Hydra. In 1739 an unknown Swiss tutor, Abram Trembley, discovered these organisms attached to plants growing in garden ditches. After a few days he discovered that when these organisms were cut, they regenerated missing parts. This discovery provoked a stir in the whole of learned Europe,[33] and before reaching Blumenbach's attention, Hydra was already a model organism supporting the epigenesist doctrine.[34]

Blumenbach framed his notion of *bildungstrieb* (formative force) around the properties of this organism, and he – as other naturalists before him – adopted it not only as a model to support epigenesis, but also as an example disproving mechanistic explanations in biology. An organism in which any part can produce the whole and in which the whole drives and shapes its parts (as shown by the celebrated experiment of turning the Hydra inside out) was hardly reducible to a mechanistic framework. As Kant wrote in his critique mirroring Hydra's endowments:

> If we consider a material whole, according to its form, as a product of the parts with their powers and faculties of combining with one another (as well as of bringing in foreign materials), we represent to ourselves a mechanical mode of producing it. But in this way no concept emerges of a whole as purpose, whose internal possibility presupposes throughout the idea of a whole on which depends the constitution and mode of action of the parts, as we must represent to ourselves an organized body.[35]

Indeed, in the same way that Hydra's parts can reproduce a whole organism, so the whole can reproduce its missing parts, so that 'An organised product of nature is one in which every part is reciprocally purpose and means'.[36] A machine, Kant insisted, is never able to reproduce or re-form both its parts and its whole structure: 'it does not replace of itself parts of which it has been deprived'.[37]

Of course, for Blumenbach the Hydra was a model organism which typified organic development in general, and Kant probably agreed on that. Hydra, with its analogical power, demonstrated that the theory of epigenesis had to be true for the entire organic realm; Hydra's multiple reproductions mirrored a universal organic property of development which, given our limited intellect, could only be grasped through a subjective principle of reason. To Kant, though, the first to establish epigenesis on firm ground was Blumenbach himself:

> As regards this theory of Epigenesis, no one has contributed more either to its proof or to the establishment of the legitimate principles of its application than Herr Holfr. Blumenbach. In all physical explanation of these formations he starts from organised matter. That crude matter should have originally formed itself according to mechanical laws, that life should have sprung from the nature of what is lifeless, that matter should have been able to dispose itself into the form of a self-maintaining purposiveness – this he rightly declares to be contradictory to Reason. But at the same time he leaves to natural mechanism under this to us indispensable *principle* of an original *organisation*, and undeterminable but yet unmistakable element, in reference to which the faculty of matter in an organised body is called by him a *formative impulse*.[38]

Hence, Kant's theoretical views of the organism reflected the observations and studies that Wolff and Blumenbach had undertaken on reproduction, development and regeneration.[39] His idea, according to which an organism needs to be considered as an object in which the causes and effects of its formation are irreducibly entwined, reflected the phenomenon of ontogeny, animal regeneration and, by analogy, all developmental phenomena. However, I think that the extraordinary influence and success of these ideas in the subsequent natural sciences (and their related philosophical discussions) did not rely so much on the conception of an organism as an intricate web of causes and effects, but more on some interesting implications of this conception. In other words, Kant defined life by indicating which phenomena mattered for a possible definition; as a natural product, an organism is only comprehensible through a teleological approach because reproduction, development and regeneration are understandable through a teleological approach. If development is the phenomenon which best characterizes life, then the study of morphogenesis is the best way to make sense of an organism. It was an extraordinary though indirect advertisement for embryology for the coming centuries. Indeed, not surprisingly, both post-Kantian and Romantic philosophies of biology, and therefore the diverse forms of organicism in the twentieth century, deemed the dynamic phenomena related to development central for any biological investigation.

However, after Kant another element was added to the list of central phenomena in life sciences: heredity. Peter McLaughlin has argued that Kant, in his writing on anthropology and race, introduced the notion of heredity into biology.[40] But as we have seen, heredity in the post-Kantian formulations had little or nothing to do with the transmission of discrete stuff, but with forces, activities and dynamic mechanisms producing specific patterns: the forms typical of the species or race. Heredity became central because it had the potential to explain development, whereas development (and its related phenomena) was seen as the manifestation of hereditary potentialities plus environment. Kant himself speculated on these things in the context of physical anthropology. When, between 1775 and 1777, he developed the theory of *Keime* and *Anlage*, he assumed that variation among races was presumably due to some specific 'structure' and 'potencies' which could be 'awakened' or lie 'dormant' according to local environmental conditions. To him, the so-called 'generative fluid' contained a predefined set of potencies, a preformed type of structural organization which, during development, blossomed according to external conditions. Kant himself, in his 'Of the Different Human Races', specified that 'the condition of the earth (dampness or dryness), along with the food that a people commonly eat, eventually produces one hereditary distinction or stock among animals of a single line of descent or race, especially with regard to their size, the proportion of their limbs (plump or slim), and their natural dispositions'.[41] Very often, Kant continued, the colour of the skin was strictly determined from the environment in which humans dwell. For instance, the black skin of the 'negroes' had the function of dephlogisticating the blood:

> true negroes live in lands where the air, because of the thickness of the trees and the marshiness of the surroundings, is so heavily phlogisticated that, according to Lind's account, English sailors run the risk of death from this cause when they ascend the river Gambia even for a single day ... it was, therefore, a very wise arrangement of Nature so to organize the skin of the negroes that their blood, even if the lungs do not sufficiently eliminate phlogiston, is yet far more thoroughly dephlogisticated than ours. Their blood must therefore deposit a great deal of phlogiston in the ends of the arteries, so that at this place – that is to say, just under the skin – it shows through as black[42]

Kant connected, in a single discourse, heredity, environment and teleology. The skin colour of 'negroes' was the result of the chemical composition of the air. But it was also a wise functional hereditary adaptation that allowed some 'races' to survive in particular environments. We could never understand why 'negroes' have black skin with a mere physico-chemical analysis; we also need a functional explanation of *why* the skin is black.

Of course, no evolutionary trends could be derived from such a scheme; in fact, neither 'structures' nor 'potencies' varied or changed, only their contextual 'expression' changed. Adaptations (and therefore new adaptive variations) were not conceived as new characteristics acquired due to the laborious and incessant activities of the organism, but as 'capacities' already included in the original stock, which was differently activated by the local 'conditions of life'. The organic form was due to both the original and predetermined organization (*Keime* and *Anlagen*) and its interaction with the environment. The living organization was due to this invisible transmission of irreducible structures that, in concert with the environment, produced all species variation visible on the earth. For instance, the variation of birds and wheat was neither due to fixed preformed structures nor to the mere influence of the environment. As Kant clearly explained in 1775: 'In birds of the same species, destined to live in different climates, there are germs [*Keime*] for the development of an additional layer of feathers when they live in cold climates: but these germs remain undeveloped when the birds are destined to live in moderate climates. Wheat has a predetermined capacity or natural tendency gradually to produce a seed of thicker hull in a cold country'.[43] Preformed tendencies and environment interacted for the creation of organic form. Variation was the outcome of heredity, environment and the functionality of the organisms (even though functionality is a mere strategy of reason). The continuity of the species and genera, in spite of the many causes operating on them during all development, could not be understood as accidental:

> We are led to this idea by the incomprehensible constancy of species and genera in the presence of so many causes streaming in upon them and modifying their development. From which I conclude that, if varieties spring up which are without exception hereditary, these could not have been brought about by a merely accidental cause, but could only been developed, and that even for this development there must be found in nature original and purposive tendencies.[44]

As all the above quotations show, Kant did not only write philosophical treatises. He published works on, among other things, plant, animal and human geography. Through these articles Kant linked his abstract philosophical insights with more concrete findings and disciplines. In so doing, he inspired geographers, anthropologists and botanists to see the earth as a single, interconnected organism. Plants, animals and environment (or diverse micro-environments) were essentially entwined. M. Nicolson has stressed the influence of Kant on German geographers, and in particular on Alexander von Humboldt.[45] Humboldt's *Essai sur la géographie des plantes*, first published in 1805, is an extraordinary document showing how diverse data can be synthesized in one unique geographical unit: geology, vegetation, temperature, chemical constitution of the air, refraction of sunlight, human activity and many other elements formed one unique interrelated system.[46] Humboldt's natural science is a very instructive example showing how important the environment was in the post-Kantian and Romantic traditions of natural history.[47] Organisms are not only whole interconnected individuals; they are also dynamically intertwined with their environments. Indeed, as Humboldt explained in one of his physiological works echoing Kant's view, 'A secret law governs all the parts of the organism, which exists only as long as all its elements are mutually means and purpose of the whole'.[48] With the presupposition that organisms were integrated individuals, Humboldt extended his analysis to the environment. As the historian of biology Erik Nordenskiöld reminds us, Humboldt 'takes as his starting point life in its entirety, examines its various manifestations, and finally dwells on the special advantages which soil and climatic conditions offer to the vegetable world in different latitudes'. He puts forward the question: how is the shape of plants affected by these conditions of life?[49]

In conclusion, Kant provided a third way between a mystical vitalism and a naive mechanism. He argued that a living organization had to be presumed because physical sciences, with their methods and principles, cannot explain it. He offered a heuristic model of biological explanation based on teleology; he maintained that parts and the whole are both effect and cause of themselves; and finally, he stressed that living forms are strictly associated with their environment. We will see that Kant, on Wolff's and Blumenbach's shoulders, provided a set of concepts and notions which, in the nineteenth century, were found extremely convincing by new generations of biologists. Subsequently, in the twentieth century these concepts and notions, filtered through Romantic 'distortions' and interpretations, were used to solve new challenges and accommodate new discoveries. In the next chapter I will analyse some of these philosophical 'distortions'; then I will show how a 'Romanticized' Kant influenced a host of young zoologists working in diverse European and American institutions during the early twentieth century.

## Romanticizing Kant

> The world plainly resembles more an animal or a vegetable, than it does a watch or a knitting-loom.[50]
>
> Immanuel Kant

Kant's philosophy and epistemology had an enormous impact on nineteenth- and twentieth-century sciences.[51] Yet Kant's philosophical and scientific influence on the budding life sciences in nineteenth-century Germany has been particularly wide, deep and enduring.[52] As Nicholas Jardine has convincingly shown, the so-called Kant-Blumenbach programme in natural history was very successful: 'From the 1790s this programme provided a framework and inspiration for a very substantial number of natural historians, physiologists and anatomists'.[53] Indeed, as we will see, Kant's bio-philosophy opened a wide intellectual space for new discussions and new interpretations of old dichotomies and issues affecting natural history since the time of Aristotle; however, he opened the way, rather than imposing a route. As F. C. Beiser wisely reminds us, post-Kantian history was essentially characterized by a constant attempt to transcend Kant's limits and problems, not to blindly or dogmatically disseminate his ideas.[54] Kant's tradition, like many others, is studded with his disciples' treasons. It was sometimes praised, exalted and revered, then betrayed, condemned and misunderstood; nonetheless, Kant's philosophy was almost always heeded, because his influence was too preponderant, the ramifications of his work too huge, and its international dissemination too wide.[55] In all its innumerable interpretations and receptions, many of the central tenets of Kant's philosophy persisted; even Kant's closer followers – Idealists, Romantics and *Naturphilosophen* – did not reject it, but transcended, transformed and expanded its original core.[56]

However, if we expect to find a coherent and well-defined research programme through the multiple translations of Kant's philosophy, we would be easily disappointed. Kant's programme was extremely sophisticated and open to different readings. As we will note in the next sections and chapters, there was not a list of 'axioms' uncritically shared, but a dynamic network of competing conceptions, traditions, ideas and methods that struggles to find a consistent disposition. As P. R. Sloan aptly noticed, Kant's programme in natural science allowed the combination of heterogeneous and contradictory elements: 'teleological purposiveness *and* their explanation by rigorous mechanism; Cuvierian functional anatomy *and* speculations about transcendental archetypes; distinctness of species *and* the concept of the unity of form in historical derivation from a common stem'.[57] Ambiguity and inconsistency also characterized versions of Romantic biology in the twentieth century. But ambiguity and inconsistency are not always the marks of bad science; instead, they sometimes point to promising new research traditions, which I see as changing networks of inconsistent approaches and ideas.

In the last thirty years many historians have highlighted the close relation between Kant's bio-philosophy and its immediate posterity. In Timothy Lenoir's controversial book *The Strategy of Life* (1982), the connection is clearly made. As he argues:

> My principal thesis is that the development of biology in Germany during the first half of the nineteenth century was guided by a core of ideas and a program for research set forth initially during the 1790s. The clearest early formulation of those ideas is to be found in the writings of the philosopher Immanuel Kant.[58]

To Lenoir, indeed, a careful reading of the published works of some of the most important German naturalists active in the nineteenth century shows how the Kantian worldview shaped and oriented a particular research programme that diverse generations of young investigators would follow – a programme Lenoir dubs 'teleomechanism'. 'Teleomechanists' – naturalists such as Gottfried Reinhold Treviranus (1776–1837), Carl Friedrich Kielmeyer (1765–1844) and Johann Friedrich Meckel (1781–1833) – were all followed by a new generation of celebrated 'teleomechanists' such as Karl Ernst von Baer (1792–1876), Johannes Müller (1801–58), Carl Bergmann (1814–65) and Rudolph Leuckart (1822–98). Kant, Lenoir argues, first set out the teleomechanist principles: a living organization cannot be explained but always posited; life phenomena require a different kind of explanation from inorganic phenomena; and the whole has a priority over its parts, function has a priority over structure and so forth. The story Lenoir tells is not about a vague chronological development of disembodied ideas within an abstract Kantian framework. It is instead a story about flesh-and-blood individuals in constant communication with one another, a lively community of naturalists deeply involved in debating philosophical and scientific issues of their time and biologists working in concrete institutions and setting feasible agendas for promising future students (and therefore future advocates of these agendas). In short, even though we do not need to accept the whole of Lenoir's narrative, I think that the so-called 'teleomechanist' research programme is more than a fiction.

Another historian of science, R. J. Richards, in his book *The Romantic Conception of Life* (2002), highlights the Kantian influence on the *Naturphilosophen*,[59] who, as L. K. Nyhart nicely illustrates, had a significant impact on the morphological research programmes in German universities (and elsewhere too).[60] In Richards's words: 'Those scientists to whom I refer as Romantic biologists generally accepted the metaphysical and epistemological propositions of *Naturphilosophie*. They took more to heart Kant's analysis of the logical similarity between teleological judgment and aesthetic judgment, which he developed in the *Kritik der Urteilskraft*.[61] However, although Kant deeply inspired the *Romantic* way of thinking about life, his doctrine underwent important revisions.[62] It is well known that his philosophy in general, including his bio-philosophical doctrine, was critically assessed and adulterated by philosophers and naturalists (broadly conceived) such as Johann Wolfgang von Goethe (1749–1832), Georg Wilhelm Friedrich Hegel (1770–1831) and Friedrich Wilhelm Schelling (1775–1854). As Isaiah Berlin clearly puts it: 'Kant hated Romanticism. He detested every form of extravagance, fantasy ... nevertheless, he is justly regarded as one of the fathers of romanticism'.[63]

Goethe had read with great interest the works of Blumenbach and Kant.[64] In 1789 his friend F. Schiller probably introduced him to Kant's essay 'On the Use of Teleological Principles' (1788).[65] By the end of 1790, Goethe had already read the recently published *Kritik der Urteilskraft*.[66] As Lenoir reports: 'From his heavily annotated copies of Kant's *Critique of Judgment* we know that Goethe himself found these passages immensely stimulating. He later acknowledged that he owed a joyful period of his life to the ideas expressed by Kant herein'.[67] With Kant, Goethe was convinced that plant metamorphosis could only be partially understood with physics and chemistry. And yet, like Kant, he believed that life exhibited purposeful activities: Goethe's views about life phenomena are teleological throughout.[68] To him, function and structure worked in tandem within a specific environment. The laws directing the formation of structure were not enough for explaining organic form. For instance, as he showed with his studies on rodents' powerful teeth in the second part of 1824's *Morphologie*, the form of both lower and upper incisors was clearly related to the specific conditions of rodent life. For a comprehensive understanding of form's creation, the conditions of existence had to be taken into account. Kant's teleological solution, according to which there was purpose without a designer, particularly appealed to Goethe.[69] Indeed, as Goethe admitted in a polemic against mechanistic views of nature: 'When people banished teleological explanation, they robbed nature of understanding; they lacked the courage to ascribe reason to her'.[70] Thus it is not so surprising that Cassirer considered Goethe's morphology much closer to Cuvier's functionalism than to Geoffroy Saint-Hilaire's structuralist approach,[71] despite the fact that, apparently, Goethe supported Geoffroy against Cuvier in their famous debate.[72] We will see that Goethe's position was still alive in the twentieth century, especially in D'Arcy Thompson's morphological views.

Goethe deeply inspired Hegel with his investigations into plants' metamorphosis. 'In the Jena correspondence', says S. S. Hahn, 'Hegel's own botanical studies were encouraged directly by Goethe's essay on the metamorphosis and morphology of plants'.[73] Goethe's teachings deeply affected Hegel's organic vision of nature, indeed, much more than a youthful experience, as Hegel included Goethe's insights on metamorphosis and organicism in the second part of his encyclopaedia, the part dedicated to his *Naturphilosophie*.[74] Yet Hegel also developed the Goethean organicist ideas (what he dubbed 'Das Organische', i.e. organics) in his 1816 *Wissenschaft der Logik* (Science of Logic), his 1820 *Grundlinien der Philosophie des Rechts* (Elements of the Philosophy of Right) and finally, in his Berlin lectures on the philosophy of history.[75] Organicist metaphors in the whole of Hegel's philosophy are utterly pervasive.[76] Even the notions of Hegel's dialectic and contradiction are essentially metaphors coming from developmental processes as seen in plants and animals. As Hahn aptly describes the way Hegel saw development: 'The seed-leaf is seen to be the negation or "contradiction" of the seed, standing in a relation of conceptual, not formal, incompatibility to its previous form. In passing from being a seed to a seed-leaf, the plant has in effect cancelled out (negated) its former part and contains the negation or destruction of its previous form'.[77] However, it is important to keep in mind that Hegel did not use a vague metaphor of the organism for applying it to some abstract notions. What he did was connect what he observed about living phenomena and processes with a more appropriate and dynamic conceptual understanding of nature.[78] In other words, to Hegel, we should think as an organism develops. The process of philosophical understanding must be seen as true epigenesis.[79] This was precisely in agreement with Goethe's methodological views about natural sciences.[80]

Hegel, like Goethe before him, was deeply impressed by Kant's third critique. Not only did Hegel approve of Kant's definition of an organism as a self-generating, purposive whole, but he was also impressed with Kant's analogy between life and aesthetics.[81] However, Hegel extended the application of Kant's conception of *Naturzwecke*. The world, reality and thought had, in Hegel's absolute idealism, the mark of the organism. From the nineteenth century onwards, Kant's organicism – as widely extended and translated by Hegel and some fellow Romantics – was 'normal science'.[82] Many *Naturphilosophen* simply assumed it was an orthodox 'paradigm'. What is probably even more interesting is that such an organicist approach was intended, by many of his advocates, as a naturalistic alternative to mechanistic philosophies.[83] In other words, no spiritual or transcendent forces had to be invoked. Nature was still ruled by laws and not by whimsical or miraculous events. Already during the very first decades of the nineteenth century, organicist philosophies *à la* Hegel tried to find a middle ground between eighteenth-century mechanisms and vitalisms, a diplomatic compromise based on a new conception of matter – a conception that Kant had already drafted in his influential 1786 book *Metaphysische Anfangsgründe der Naturwissenschaft*.[84] Gode-von Aesch recognized long ago that most of the Romantic naturalists abhorred the idea of the vital force as detached from matter. To them, there was not a dichotomy between a living organization and inert matter insofar as inorganic phenomena were nothing but a special case of an animated universe: 'a major contribution of romantic thought was precisely that it recast all sorts of dualistic conceptions in terms of a biocentric universalism'.[85] Hegel not only inherited such a biocentrism from previous naturalists and philosophers, he also widely publicized it through his authority and fame. Indeed, as is well known, Hegel's philosophy was hugely influential, even in the twentieth century. His philosophy was studied not only by philosophers, historians, politicians and social theories, but also by biologists and physiologists. J. S. Haldane is a striking example.[86]

Less influential than Hegel, Schelling too proposed a biocentric philosophy. He was a friend of Hegel from the years of Tübinger Stift, a Protestant institution that both attended in the 1790s. A scholar of Kant, Schelling was also deeply inspired by Erasmus Darwin, Blumenbach, Goethe and Kielmeyer.[87] With Hegel, he was fascinated by the distinctiveness of life phenomena. In 1802 he received the title of 'Doctor Medicinae' from the Bamberg medical school.[88] As he admitted four years later in the *Annals of Medical Science* (1806): 'The science of medicine is the crown and blossom of all the natural sciences, just as the organism in general, and the human organism in particular, is the crown and blossom of the world'.[89] Drawing on Kant's third critique, Schelling proposed one of the most ambitious organicist philosophies. He saw all of nature, the entire cosmos, as a developing organism. His organicist metaphysics intended to resolve Kant's dichotomies between the subjective and objective, mental and physical, phenomenon and noumenon. Schelling, like most Romantic naturalists and philosophers, also meant to go beyond Kant's regulative limits: it is not that we need purpose as an epistemic strategy in order to understand the organism; to Schelling, the entire cosmos, the reality, *is* in fact an organism. As Beiser explains: 'if nature is an organism, then it follows that there is no distinction in kind but only one of degree between the mental and the physical, the subjective and objective, the ideal and the real. They are then simply different degrees of organisation and development of a single living force, which is found everywhere within nature'.[90] Essentially, Schelling linked two central elements of Kant's late philosophy: the idea that matter is not inert but active, and nature's unity and harmony.

Although Schelling's philosophy did not share the fame of Hegel's system, his *Naturphilosophie* was quite influential among young physicians, embryologists and naturalists. For example, as von Baer remembers in his autobiography, Schelling's *Naturphilosophie* was widespread among nineteenth-century naturalists: 'I was very eager to take a consistent course of lectures on the philosophy of Schelling. Nature philosophy was spoken of everywhere and mentioned in many books'.[91] Von Baer adds that his advisor in Wurzburg, Ignaz Döllinger, was not only an enthusiast of Kant's philosophy but also had personal contact with Schelling: 'he had studied the philosophy of Kant with zeal and then he was carried away by Schelling with whom he had personal contact'.[92] Like von Baer, Döllinger trained in physiology and anatomy with L. Agassiz, C. H. von Pander and L. Oken, all influential naturalists. Even though later in their lives Döllinger and von Baer became rather critical of the speculative excesses of nature philosophy, for a period of their life they surely were attracted.[93] Whatever the case, von Baer demonstrates that for many naturalists active during the first decades of the nineteenth century, Schelling's philosophy could not be overlooked so easily. Praised or criticized, Schelling's *Naturphilosophie* was not ignored: it addressed philosophical issues that deeply interested those naturalists studying life phenomena.

Another fundamental dichotomy that Schelling's philosophy aimed to overcome was the opposition between mechanism and vitalism. Like Hegel and most of *Naturphilosophen*, Schelling despised both. To him, the right form of scientific explanation should start from the fact that nature is intrinsically dynamic, organic and creative. As Joseph Esposito remembers: 'in the life sciences [Schelling] found congenial arguments that justified a holistic approach to organisms, but that were also based on a dynamic evolutionary model and not on the acceptance of specific vital entities like souls or entelechies'.[94] To Schelling, the dichotomy between vitalism and mechanism was based on the false assumption that organization followed from inert matter, from a dead universe. However, if we consider a living organization as primordial, as the main ontological element structuring reality, we can reconcile the two positions. As Schelling explained in his 1796's *Ideen zu einer Philosophie der Natur* (Ideas for a Philosophy of Nature): 'all organic product is for itself, his existence is not dependent on any other entity ... The organization produces itself, arises from himself ... Each organic product is the reason of his existence in itself, for it is by itself cause and effect. No individual part could arise outside the whole, and this whole itself exists only in the interaction of its parts'.[95] Nature was, for Schelling, the result of dynamic and opposite forces. Nature was a living organism that could never been understood through the inert category of mechanism. Nature was not something given once for all; it developed and changed constantly. The aim of the *Naturphilosophen* was therefore to understand the movement of nature in its *genesis*, its incessant and creative becoming. M. L. Heuser Kessler, A. Leyte and Esposito have connected such an organicist position – in particular Schelling's attempt to resolve the dichotomy of mechanism versus vitalism and his ideas about active matter – to some contemporary ideas in physical sciences and general system theory. According to them, Schelling should be seen as a precursor of von Bertalanffy, I. Prigogine and H. Haken in terms of their conception of self-organization because, like them, Schelling aimed for a 'physics of becoming' against the 'physics of being'.[96] To Schelling, indeed, nature is defined as pure productivity, organic activity and process: 'Things are therefore not principles of the organism, but rather conversely, the organism is the principle of things', he says in 1798's *Von der Weltseele* (On the World-Soul).[97]

## Beyond the German-Speaking World

The 'panorganic' philosophies of Goethe, Hegel and Schelling did not die with them. Their form of organicism widely influenced later nineteenth-century natural philosophers and, especially, the first naturalists called 'biologists'. As Esposito comments: 'it is often forgotten that the tradition of speculative *naturphilosophie* in Germany did not altogether die with the passing of Schelling and Hegel, but continued up to the present century'.[98] Yet outside Germany, the philosophy of both Schelling and Hegel travelled to Britain and the United States. In Britain, philosophers such as T. H. Green, F. H. Bradley and J. M. McTaggart continued and extended the teachings of Hegel. As we will see in Chapter 3, the movement of British idealism was not only influential among philosophers, but it also inspired scientists. A similar situation occurred in the United States. As G. Stanley Hall, a physiologist and psychologist who studied under W. James, recorded in 1879 about the diffusion of evolutionary idealism in the US: 'the influence of German modes of thought in America is very great and is probably increasing ... the market of German books in the United States is in several departments of learning larger than in Germany itself'.[99] One of the most popular American philosophers, C. S. Peirce, was deeply impressed by Schelling's *Philosophie der Natur*. As he wrote in a letter to James: 'I consider Schelling as enormous ... if you were to call my philosophy Schellingism transformed in the light of modern physics, I should not take it hard'.[100] As we will see in Chapters 5 and 6, 'German modes of thought' also entered zoology departments and laboratories. It was behind a new form of twentieth-century organicist biology. However, from Kant onwards, what kind of organicism was really bequeathed to the biologists in the twentieth century? In other words, which elements, ideas and conceptions were retained and which were discarded or simply forgotten? Is there an intellectual continuity or a mere similarity of thought? We need a brief digression.

In 1986 Lenoir argued that between the Kantian teleological conception of organic phenomena and the *Naturphilosophie*'s approach there was a profound difference. In particular, Lenoir drew a sharp line between 'teleomechanists', who developed a successful empirical programme based heavily on Kantian philosophy, and the Romantic *Naturphilosophen*, who set out a speculative, vague and abstract agenda that was far less successful. Lenoir had very good reasons (both historical and conceptual) to draw such a distinction; many figures he labelled 'teleomechanists' criticized the mystical excesses of *Naturphilosophen* (for example, von Baer, Cuvier and Müller). However, the nature of such a distinction is not always easy to support because, as Nyhart nicely put it, 'Most early nineteenth-century writers on form confound categorization schemes based on rigid philosophical distinctions, they appropriated the language of Kant, of Schelling, and of Cuvier in different places'.[101] In fact, more recent historiography has stressed the continuity, the strong intellectual relations and the evident similarities between Kant's bio-philosophy and the Romantic conception of life. In other words, what seems to be emerging is a more variegated and sophisticated picture. Post-Kantian morphologists, physiologists, embryologists or zoologists and naturalists absorbed Kant's and Romanticism's teachings in their own ways and according their contextual needs. As Beiser has finally commented:

> it seems to me inadvisable to make a sharp and fast distinction between *Naturphilosophie* and the tradition of German physiology and biology, as if *Naturphilosophie* were a corrupt metaphysics flaunting Kant's regulative guidelines, and as if physiology and biology were hard empirical sciences heeding them. Ultimately, there is only a distinction in degree, and not in kind, between Schelling, Hegel, and Novalis on the one hand and Bluemenbach, Kielmeyer, and Humboldt on the other.[102]

One of the most important transformations of Kant's philosophy was the reformulation of his reflective judgement. Kant had famously stated that 'there will never be a Newton for a blade of grass' because no determinant judgement could ever be consistently applied to the organic world. In making this statement, Kant assumed that no life science could ever reach the status of physical science, which was considered the ideal model of reliable knowledge. To many of Kant's followers, the reflective judgement imposed too harsh a limit on the study of life because it could only provide inductive generalizations grouped together by the mere subjective heuristic of reason; in other words, there could never be a proper science of life. Post-Kantian naturalists, including transcendental morphologists and philosophers of nature, felt the need to extend or go beyond Kant's distinction; they conflated reflective and determinant judgement and 'ontologized' or naturalized purpose in nature. By then, teleology was not a mere strategy of reason used as a tool to interpret organic phenomena, but was a constitutive or inherent property of organisms themselves (and often, of the whole of reality). In other words, the student of life requires a teleological approach not as epistemic heuristic justified by the limits of his/her intellect, but because life *is* intrinsically purposive. We will see that not only was the distinction between reflective and determinant judgement reframed or forgotten in the post-Kantian tradition, but it was also simply ignored in twentieth-century organismic philosophies.

If we consider traditions not as monolithic and static sets of ideas, beliefs or practices coming from a unique source,[103] but as a result of a dynamic and complex sedimentation of elastic notions coming from different contexts, the distinction between teleomechanists and *Naturphilosophen*, or between *proper* neo-Kantians and Romantics, does not need to be emphasized. What should be emphasized, though, is how Kantian organicism was transformed and translated. Indeed, notwithstanding Lenoir's distinction, it seems quite difficult to question the influence, direct or indirect, of Kant's Romanticized bio-philosophy on disparate research agendas set out in Europe throughout the nineteenth century.[104] Lenoir's teleomechanist programme, the morphological tradition that Nyhart portrays and links to German universities, the Romantic conception of life that Richards depicts in his large volume, and the historical reconstructions of Beiser, Humenan and a few others all exhibit a similar, recurring way of thinking. Yet such a broad 'way of thinking', as I mentioned earlier, was not extinguished in the nineteenth century, nor did it remain confined to Germany. It was exported to other countries where it was critically assessed, slowly metabolized and, consequently, adapted to new contexts and needs.

Lenoir never contemplated the possibility that such a tradition, or way of thinking, would survive the nineteenth century and even be exported outside the German-speaking world. To him, the 'teleomechanist' research programme died with von Baer, Bergmann, Bischoff, Virchow and Leuckart. But as I will show in the next sections, it was not so. Indeed, some forms of post-Kantian Romantic thinking filtered into England and thrived during both the nineteenth and twentieth centuries, informing some of the most interesting debates in biosciences during the first decades of the twentieth century. Yet towards the end of the nineteenth century and the beginning of the twentieth, we find important clusters of 'neo-Kantianism' and Romantic biology in the United States too. In fact, the figure considered by Lenoir the last 'teleomechanist' in Germany, Rudolf Leuckart, trained some of the most interesting twentieth-century American biologists – biologists who, once back in their own country, did not reject the teachings, approaches and ideas they had learned abroad.[105]

Other authors, as I mentioned, hinted at an intellectual relation between Romantic views and twentieth-century system biology (Cassirer, Lamb, Leyte, Müller, Heuser Kessler and Esposito, to name just a few). However, nobody really displayed clearly the network of reciprocal influences or the conceptual and institutional landscape that made possible the survival of a Romanticized Kant well into the twentieth century. In other words, they saw that there was an intellectual similarity, but they did not know how it came about. A mysterious gap lies between eighteenth- and nineteenth-century organicism and twentieth-century organismal biology (including the London congress in 1931). The aim of the next chapter is to fill this gap.

# 2 ROMANTIC BIOLOGY: ESTABLISHING CONNECTIONS IN THE NINETEENTH AND TWENTIETH CENTURIES

> Every organized being forms a whole, a unique and closed system, whose parts mutually correspond and concur to the same definite action by a reciprocal reaction.[1]
>
> Georges Cuvier

Kant's organism, and its various romantic variations, soon spread in other countries. In England, Kant's critical tradition thrived through complex networks. After the 1830s Kant's whole philosophy was widely discussed,[2] especially thanks to the authority of William Whewell (1794–1866) and his 1837 *History of the Inductive Sciences*. For instance, Whewell's defence of Cuvier's functional biology was based on Kant's organicism.[3] Siding against the French naturalist Geoffroy Saint-Hilaire, Whewell noted that 'Whether we judge from the arguments, the results, the practices of physiologists, their speculative opinions, or those of the philosophers of a wider field, we are led to the same conviction, that in the organized world we may and must adopt the belief that organization exists for its purpose, and that apprehension of the purpose may guide us in seeing the meaning of organization'.[4] As a famous Cambridge philosopher of science during the first decades of the nineteenth century, Whewell was in a very effective position for influencing new generations of British naturalists, including Owen and Darwin.[5] However, Whewell was not alone. P. Rehbock has showed how Kant's ideas, filtered through Goethe, Schelling and some celebrated *Naturphilosophen*, were absorbed and transformed by other British philosophers and naturalists.[6] The poet Samuel T. Coleridge (1772–1834), the anatomist Robert Knox (1791–1862) – who deemed Kant, Goethe and Oken individuals belonging to a race superior even to the Anglo-Saxons[7] – the embryologist Martin Barry (1802–55), the physician Joseph H. Green (1791–1863), the physiologist William B. Carpenter (1813–85), the comparative anatomist Richard Owen (1804–92) and the physician John Goodsir (1814–67) all imported and transformed strands of German bio-philosophy in the United Kingdom. As Nicolaas

Rupke notes, 'Knox, Grant, Barry, Carpenter, Green and several lesser figures all sought to reform the biomedical sciences by borrowing the Romantic, or transcendental, program from continental sources'.[8]

Among the British naturalists, the main vehicles through which Kant's organicism filtered were von Baer and Cuvier. Barry was first responsible for the introduction of von Baer's biology in the UK.[9] In 1853 Thomas Huxley translated part of von Baer's *Ueber Entwicklungsgeschichte der Thiere* (Fragments Related to Philosophical Zoology): 'it seemed a pity', Huxley said when introducing von Baer's text, 'that works which embody the deepest and soundest philosophy of zoology, and indeed of biology generally, which has yet been given to the world, should be longer unknown in this country'.[10] Yet various scientific and literary journals provided translations and commentaries that introduced British naturalists to the difficult ideas coming from Germany and France. On the continent, the intellectual link between Kant and von Baer's embryology was first shown by E. S Russell in 1930, when he wrote that 'von Baer's philosophical standpoint is definitely anti-materialistic. He is clearly influenced by Kant's teleology'.[11] More recently, Lenoir extended the relation between Kantian philosophy of biology and von Baer. Indeed, for Lenoir, von Baer improved the vitalo-materialist programme first set out by Kant and Blumenbach – an agenda that was essentially organismal and anti-reductionist. As Lenoir put it: 'I would urge that von Baer's work brings the conceptual framework enunciated by Kant and Blumenbach in their writings of the late 1780s and 1790s to its most robust formulation ... not only was von Baer conversant with the works of these men; he built their leading concepts into his own theories of the developmental schema and primitive organs'.[12] Von Baer's conception of the organism was essentially Kantian. In his autobiography, he remembered how important it was to consider the relationships among parts in one individual organism: 'an insight into the structure of an animal form is possible only when one can survey all the essential components of organisation in their mutual relationship'.[13] And yet, as he himself wrote in his celebrated monograph *Entwicklungsgeschichte* in 1828, showing his clear debt to Kant and Blumenbach:

> But all explanations of this [materialistic] kind the physiologist finds soon to be highly incomplete, since they touch only one single side of life; and he comes to see, above everything, that life cannot be explained from something else, but must be conceived and understood in itself. The time is approaching when even the physicist must admit that in his investigations he merely put together the isolated physical antecedents of the totality of life, and thereby fashions for himself an artificial beginning.[14]

If von Baer was one of the most successful propagandists of Kantian organicism, he was not alone. Indeed, Russell, in his classic *Form and Function* (1916), stressed the similarities between the von Baerian approach to biology and that of Cuvier; both endorsed a functional biology and both recognized the importance of embryology in order to establish homologies,[15] but, especially, both were Kant's pupils. Cuvier, Russell claims, was a 'teleologist after the fashion of Kant, and there can be no doubts that he was influenced, at least in the exposition of his ideas, by Kant's *Kritik der Urtheilskraft*, which appeared ten years before the publication of the *Lésons d'Anatomie Comparée*. Teleology in Kant's sense is and will always be a necessary postulate in biology'.[16] Russell's historical argument was extended by Cassirer, who in 1950 saw a deep affinity between Kant's organicist agenda and Cuvier's research programme. Cassirer argued that Cuvier's idea of the type, and even his principles, i.e. correlation of parts and conditions of existence, were corollaries of Kant's definition of an organism as an interconnected system.[17] In other words, the empirical and theoretical tenets of Cuvierian biology were well rooted in Kantian speculation on the nature of the organic beings. This is not surprising. Cuvier knew the German context well. Between 1784 and 1788 he spent part of his training in Germany. Bilingual in German and French, he had been a student of Karl Kielmeyer at the Karlsschule in Stuttgart and was familiar with the works of Blumenbach.[18]

Further evidence demonstrating that Cuvier was well aware of Kant's bio-philosophy comes from his own writings, as the following quotation taken from his *Lésons d'Anatomie Comparée* shows:

> This general and common impulsion of all the elements [of a living body], is to such an extent the very essence of life, that parts which are separated from a living body quickly die, because they themselves do not possess their own impulsion, and only participate in the general movement which guarantees their union. In this way, as Kant has pointed out, each part of a living body is as it is, because of its work towards the whole; while in the case of inorganic bodies, each part exists for and by itself.[19]

In addition, at the end of the eighteenth century there were important connections between German and French science, and these connections were fostered by political factors. Indeed, the Napoleonic politics of expansion eventually encouraged, indirectly, the diffusion of Kant's bio-philosophy in France. As Lenoir states:

> In the 1790s all roads led to Gottingen, where the young men in Blumenbach's inner circle were envisioning a comprehensive approach to organic nature. Between 1800 and 1815, however, particularly after the German states had largely become satellites of France and many German universities were closed, those roads led to Paris, which for German zoologists meant they led to Georges Cuvier. During these years, when Alexander von Humboldt made Paris his home base for exploring the world, a small German colony sprang up around Cuvier. These young Germans had several features common in their backgrounds. They were either students of Blumenbach and Reil, or, through contact with their students, were about to become enthusiastic converts to the teleomechanist program.[20]

Kant's philosophy of biology – his stress on function rather than structure, the prominence of the whole over its parts, and teleology as a necessary heuristic strategy to understand living beings – ran deep in the theoretical convictions of von Baer's embryology and Cuvier's functional biology. Thus the influence of von Baer's and Cuvier's biological thought spread, directly or indirectly, across all of nineteenth-century continental Europe.[21]

As we have seen, during the first half of the nineteenth century in England there were diverse key figures who adopted the agenda of a functionalist and organismal philosophy of biology. I will only mention in detail two of them: Joseph Henry Green and the controversial anatomist Richard Owen, who in 1831 went to Paris to attend Cuvier's lectures.[22] Green was a British surgeon and comparative anatomist. He studied in Germany as a medical student in Berlin and Hannover and, once back in England, he became a close friend of the poet and philosopher Coleridge, who, in turn, was an enthusiastic advocate of German Romantic philosophy.[23] As Sloan vividly describes Green's science:

> Green's combination of medical expertise, his personal study of German philosophical and scientific tradition, and his professional appointments all made him an ideal vehicle by which a dynamic appropriation of German philosophical biology was realized ... Green's philosophical synthesis of Kant, Schelling, Solger, and Blumenbach with endemic British philosophical, biological, and medical thought represents the most creative assimilation of the German tradition I have located among the British at this time.[24]

Even though Green may have been influential, especially through his very successful lectures held at the College of Surgeons in London for five spring terms (between 1824 and 1828), the figure that most inspired British naturalists was Owen, the British Cuvier. Born in Lancaster in 1804, Owen was trained as a physician in Edinburgh and at St Bartholomew's Hospital of London. In 1826 he attended Green's lectures and was deeply impressed. Although Green introduced Owen to German science, he was also profoundly stimulated by Whewell's philosophy,[25] Goethe's morphology and Oken's and Carus's *Naturphilosophie*.[26] Like Kant, he defined organisms as indivisible bodies: 'we may say that organized bodies are distinguished from unorganized by their indivisibility – by their incapacity to suffer division with a continuance of their characteristic properties in the separated portions.' In the inorganic world, Owen continued, 'the adaptation of the different parts to the Scope of the whole is wanting'.[27]

From the 1830s onwards, Owen began to find an eclectic synthesis between Cuvier's functionalism and Geoffroy's structuralism.[28] To him, functions could not always explain form because very different structures were often employed by organisms for performing the same function. After all, as Goethe had already noted in his *Die Metamorphose der Tiere* (1799–1800), form and function had to be considered as complementary phases in one unique developmental process.[29] Owen, together with many others of his fellow naturalists, followed such guidance. But Goethe's and Owen's solution to the dichotomy between form and function was not unique. The polarization between functional and structural biology – a polarization mainly diffused ever since the famous debate between Cuvier and Geoffroy, and highlighted by Russell in his first influential book *Form and Function* – was not the norm. In fact, the very neat division between pure structural biologist and pure functional naturalist is more a historical fiction than a true distinction. As we will see, hybrid forms that involved both structure and function were more common, even during the first decades of the twentieth century. Owen's later idiosyncratic beliefs about evolution – seen as a teleological process driven by sudden 'transformations' already preordered in the 'archetype' – already reflected a sophisticated hybrid of structural and functional considerations. The vision of Owen, as the vision of many of his contemporaries, was a synthesis of different stances. Cuvier's functionalism and Geoffroy's transcendentalism were unified at a superior, cosmological level. As Sloan recorded: 'Owen's teleological view of nature was in many respects that of Kant and his interpreters, an immanent teleology that manifested itself in a complex historical process. Nature is a system of natural forces, but a system working for the realization of an immanent goal, the goal of human consciousness and moral freedom'.[30]

Even though the sources on which those naturalists and thinkers drew could be heterogeneous, and although von Baer and Cuvier certainly exerted their direct or indirect influence, there was another towering figure impacting British science during the first half of the nineteenth century: Johannes Müller. In 1833 Müller published an influential book, *Handbuch der Physiologie des Menschen*, which was translated in England as *Elements of Physiology* in 1842 and, as Ospvat recounts, was 'widely used as a textbook throughout Europe'.[31] Müller was also an enthusiastic follower of von Baer and Kantian philosophy. And, as Adrian Desmond shows, *Elements of Physiology* influenced Owen deeply, as well as many British scholars fighting against the new materialist biology and morphology inspired by the French anatomist Geoffroy Saint-Hilaire. In fact, as Desmond puts forward in his *The Politics of Evolution* (1989), such an anti-materialist tradition, well characterized by von Baer, Cuvier and Müller, and then by Green, Barry and Owen, represented in England a bulwark not only against the 'subversive' biology of Geoffrey but also against the materialist doctrines of transformationism (evolution) led by various Lamarckians and radical reformers. In short, throughout the nineteenth century, organismic and teleological biology represented the conservative orthodoxy in Britain – an orthodoxy that was never really upset by various reformers who advocated for a materialist, transformist, anti-teleological biology. Kant's bio-philosophy, filtering through the rhetoric of influential advocates belonging to the British nineteenth-century establishment, left an important legacy to the next generations of biologists. However, between the late nineteenth and early twentieth centuries, Kant's bio-philosophy stopped being an orthodox tradition. With the triumph of Darwin's biology on one hand, and other forms of Lamarckian materialism on the other, Kant's bio-philosophy increasingly fell into the background. In Britain and United States, at least, it became a philosophy of minorities and began to be associated with leftist biologists. We will see in the next section that the British neo-Kantian tradition advocated by figures such as J. S. Haldane, D'Arcy Thompson, E. S. Russell and H. J. Woodger, although owing its general theoretical framework to the past, did not derive smoothly from the nineteenth-century orthodoxy and its establishment. Its route had been much more complicated and interesting.

## Forms of Organicism in Twentieth-Century Britain

In the previous section I have sketched the multiple possible routes through which Kant's organicist tradition travelled in England during the nineteenth century. However, if we expect a direct line between nineteenth-century British embryologists and anatomists and twentieth-century British biologists, we will soon be disappointed. Indeed, when we consider the new generation of 'Romantic' biologists born after the second half of the nineteenth century, and then well established in twentieth-century biology, we observe that they had few things in common with Owen, Coleridge, Green and Barry. Surprisingly, their direct sources were not their fellow older countrymen, but essentially the late nineteenth- and twentieth-century German biologists. In other words, whereas the earlier nineteenth-century enthusiasts of Kant's bio-philosophy certainly spread the myth of the superiority and quality of German biology in England, they did not influence, apart from incidentally, the new generations. As I will show in a moment, these new generations adopted Kant's Romanticized bio-philosophy either directly in Germany or through German sources. What I think the previous British anatomists did achieve, though, was to convince the whole scientific establishment that German science generally mattered. During the late nineteenth century, if one wanted to be a good biologist, one had to study German.

Because there is not a unique and direct line drawing a map of reciprocal influences, if we want to get a general idea of how some leading British bio-scientists could have been influenced by Kant and the Romantic organicism during the late nineteenth and the twentieth centuries, we need to look at individual biographies which will show a kaleidoscopic geography of intellectual and institutional connections. Take the Scottish physiologist J. S. Haldane (1860–1936). According to Martin Goodman,[32] two key figures influenced Haldane's biological views: the English-German physiologist William Thierry Preyer (1841–97) and the German botanist Eduard Strasburger (1844–1912). Although Preyer was born in Manchester, his training was mainly in German countries, attending university in Bonn, Berlin, Vienna and Heidelberg. Graduating in physiology at the University of Heidelberg, Preyer studied with Max Schultze (1825–74) in Bonn[33] and, later, with Claude Bernard (1813–78) in France.[34] Given the context in which Preyer undertook his training, it is not surprising that he, in his influential *Elemente der allgemeinen Physiologie* (1883; translated into French in 1884), endorsed an organicist position about the study of life. In fact, Preyer's teacher at Bonn, Schultze – mainly remembered as one of the co-founders of cell theory – had been a pupil of Johannes Müller in Berlin. Furthermore, the Romantic 'discourse' that Preyer put forward was particularly evident when he discussed the relation between physiology and other sciences, and also when he tried to shed light on the concept of life. Firstly, Preyer stressed the importance of physics and chemistry for the study of physiology – a science he defined as 'the pure science of the life's functions'.[35] Physiology remained an independent science, nonetheless, because:

> As explicative science, the theoretical physiology belongs to the exact sciences, such as physics. However, physiology is different insofar it is above all an applied science, a physics, a chemistry, an applied morphology, whose the results are used for the study of the physiological phenomena. In the natural phenomena, the pure physics only regards the transformations of forces, pure chemistry the changing of the matter, pure morphology the modification of forms: the pure physiology study the natural phenomena that, as in the living bodies, presents all the three orders of changing.[36]

Preyer's understanding of the concept of life is even more explicit in the following quotation:

> The real nature of all the series of movements we call life is the result of facts of which physics and chemistry are not concerned because, without a transformation of their principles, they cannot explain life. Both sciences, physics and chemistry, are only concerned about the physical and chemical properties of the bodies; now the psychical and the phenomena of evolution do not belong to such a domain ... they are the object of physiology.[37]

The methods used by physicists or chemists, though useful, cannot be considered sufficient for the study of life's functions. Haldane, as we will see, was deeply influenced by these ideas.

Strasburger held similar convictions to Preyer. Born in Warsaw in 1844, he studied in France and Germany, obtaining his PhD at the University of Bonn in 1866. In 1894 he published the famous and still-used *Lehrbuch der Botanik* (Textbook of Botany), which went through thirty-five editions and was translated into eight languages (the last edition was published in 2002).[38] In the first part of his monumental work he dedicated a section to the 'physical and vital attributes of plants', a part in which he clearly exposed some organicist ideas informing botanical research. Plants, Strasburger argues, are 'of the nature of solid bodies'.[39] Therefore they have a weight and density; they have a conductivity of light, sound and electricity. However, he concluded, even though plants share many physical properties with inorganic matter, material and chemical substances are not enough to characterize life in itself. In Strasburger's own words:

> No other substance exhibits a similar series of remarkable and varied phenomena, such as we may compare with the attributes of life. As both physics and chemistry have been restricted to the investigation of lifeless bodies, any attempt to explain vital phenomena solely by chemical and physical laws could only be induced by a false conception of their real significance, and must lead to fruitless results. The physical attributes of air, water, and of the glasses and metals made use of in physical apparatus, can never explain qualities like nutrition, respiration, growth, irritability and reproduction.[40]

The training Haldane received in Germany, as Goodman has emphasized, had been decisive for the development of his thought.[41] Once back in Britain, he continued to expand his knowledge by studying Hegel's philosophy, especially through the filter of the Scottish idealist tradition and T. H. Green (1839–82), an Oxford neo-Hegelian philosopher and one of the most prominent figures of the British idealist movement.[42] Yet Haldane also had occasion to discuss and deepen his biomedical convictions with D'Arcy Thompson,[43] a lifelong friend. The two were born on the same day of the same year, and both were Scottish and Edinburgh students. In 1884 Haldane was hired as a demonstrator in physiology at the University College of Dundee, and there he worked with Thompson in checking the quality of air breathed in slum houses occupied by miners in nearby Dundee.[44] Although there were many similarities between Haldane's and Thompson's training, Thompson was particularly influenced by the embryologist Francis Maitland Balfour (1851–92) during his stay in Cambridge. Balfour started his career very early: in 1874 he was one of the first scientists working at the Naples Zoological Station. And in the same year, while still an undergraduate student, he published with Michael Foster (an eminent Cambridge physiologist who spent part of his training in France and Germany[45]) the influential textbook *The Elements of Embryology* (1874), a book whose introduction praised von Baerian embryology: 'the advances made in Vertebrate Embryology, through the elaborate work of Remak, the labors of Rathke, Allen Thomson and others, the admirable lectures of Kolliker, and the researches of more recent inquirers, though many and varied, cannot be said to constitute any epochs in the history of the subjects, such as that which was marked by Von Baer'.[46] Balfour was a central figure for British biological sciences; as director of the Cambridge Morphological Laboratory, he attracted students such as W. Bateson, W. Weldon, A. Sedgwick and, of course, D'Arcy Thompson.[47] It is probable that through university textbooks such as those of Foster and Balfour (in 1880 Balfour also published a large two-volume treatise on comparative embryology), a new generation of investigators absorbed the continental embryological tradition. In fact, as Blackman comments, 'the book [*The Elements of Embryology*] provided a standardized routine for students, taking them step by step through the laboratory work required for part of the course in elementary biology'.[48]

In the biography written by D'Arcy Thompson's daughter, Balfour's importance in the development of Thompson's conception of biology is stressed over and over again. As she wrote: 'the man who came to mean most to him, whom he loved and revered, who was his guide, philosopher, and friend, was Frank Balfour'.[49] Yet as Thompson said during his inaugural lecture at the University of Dundee in 1885:

> [as] a pupil under F. M. Balfour, I learned the power of a really great teacher, which is a rarer thing even than a great discoverer ... under Balfour ... one great branch of biology actually grew up in his end into science – the science of embryology. He gathered up all scattered knowledge that was hidden in books and floating in men's brain, and he, for the first time, wove it together, and entwined with it the thread of his own originality and genius.[50]

In order to give a broader glimpse of the Cambridge Morphological Laboratory and its strong continental ties, we may consider another influential figure: Balfour's pupil and former demonstrator, Adam Sedgwick (1854–1913). Sedgwick graduated from Cambridge in 1891, in natural sciences, and during his training he was deeply influenced, as Marsha Richmond recounts, by Michael Foster and Frank Balfour.[51] Sedgwick was an influential teacher, and his views 'found a following among his students and future leading figures in British Zoology, including Gavin de Beer, C. Clifford Dobell, James Gray, E. W. MacBride, E. S. Russell, and D'Arcy Thompson'.[52] Looking at the introduction to Sedgwick's *A Student's Text-Book of Zoology* (1898), we find that Sedgwick clearly endorsed some central tenets of Romantic bio-philosophy. Although the student of life, the biologist, should never presuppose the existence of a vital force – 'A vital element, i.e., an element peculiar to organisms no more exists than does a vital force working independently of natural and material processes'[53] – he could never explain life on the basis of physics alone. In fact, life could never be reduced to physico-chemical structures and compounds, because even though the 'properties and changes of living bodies are strictly dependant on the physico-chemical laws of matter', the natural conditions under which living organisms work 'distinguish organisms from all organized body'.[54] It is the peculiar organization of structures which makes life, not a mere physico-chemical substrate. Finally, Sedgwick concluded, even though we are unable to find any substantial difference between inorganic and organic bodies, in observing even the simplest organisms, we cannot regard life as the result of the mere and simple movement of matter. The biologist, if he/she wants to explain living phenomena, must presuppose the existence of a living organization insofar as we are unable to get any further reduction. The existence of living bodies 'presupposes, according to our experience, the existence of like or at least very similar beings from which they have originated'.[55]

E. S. Russell belonged to a younger generation of scholars. He studied at the University of Glasgow in the early twentieth century, and after his graduation he went to work in Aberdeen with J. A. Thomson (1861–1933), a Scottish naturalist who in 1899 was appointed Regius Professor of Natural History at the University of Aberdeen. Thomson was a former student of Haeckel in Germany and a close friend and colleague of Patrick Geddes, with whom he published several books.[56] A prolific writer, Thomson endorsed an anti-mechanistic conception of life, a conception elaborated through the study of Bergson and the direct influence of Geddes.[57] He, with Sedgwick, accepted a form of organicism according to which, though life depended on matter, it was irreducible to physico-chemical properties. In our possible experience of living organisms, he argued, life always comes from life, and we have no evidence of how life has been built from inorganic matter: 'life probably began when the conditions of heat and solubility of substance were more favourable to the formation of peculiar and complex matter than at present. But such a statement is often thought to be unphilosophical in view of the fact that we have at present no experience of the formation of such substances, and that it has been conclusively proved that living creatures always proceed from preexisting life'.[58] Once again, the biologist should presuppose the existence of life rather than reduce it to simpler inorganic elements. As Thomson would state again during his Gifford lectures in 1915–16: 'We considered the organism under the category of a material system ... we reached the conclusion that, while this is a legitimate and useful way of looking at a living creature, the formulae of physics and chemistry are inadequate for the re-description of the everyday bodily functions, or of behaviour, or of development, or of evolution'.[59]

If Russell was indebted to the bio-philosophy of Thomson, he was also influenced by Geddes. Patrick Geddes (1854–1932) had been a student of Thomas Huxley at the London School of Mines, and he spent part of his training in Europe. In 1878 he went to France, where he worked at the Marine Station in Roscoff, and a year later to Italy, where he settled at the Naples Zoological Station. A prolific and polymathic author, Geddes also endorsed an anti-reductionist approach to life sciences, and as his co-authored books with Thomson demonstrate,[60] Geddes argued against a mechanistic approach to biology.[61] As Michael Graham stresses in his short biographical sketch of E. S. Russell, it was Geddes who fostered Russell's interest in philosophical issues in biology. Indeed, all the philosophical books Russell published, Graham argues, are 'corollaries of the main, essentially Geddesian, conclusion of *Form and Function*, that no appreciation of animals has been satisfactory that has rested content with analysis into parts, whether anatomical, physiological, or Pavlovian. The meaningful entity is the "frog", not its webbed feet, nor its reflex jumping, and studies will be fruitful when they end by considering the whole frog'.[62] Russell was a trained zoologist but also a skilled historian of science, and as we will see, he underpinned and justified his organismic approach with a learned and historical reconstruction of functional, organismal and anti-mechanistic biology subsequent to Aristotle.[63]

Woodger was the youngest of the group I am considering. He was born in Norfolk in 1894 and went to University College London in 1911.[64] There he was trained by J. P. Hill (1873–1954), head of the zoology department at that time and Professor of Embryology. Hill[65] had been a student of J. Beard (1858–1923) in Edinburgh. Beard was Huxley's pupil and spent part of his training in Wurzburg and Freiburg studying embryology.[66] After World War I, Woodger worked as an assistant zoologist and comparative anatomist at University College London. The young Woodger was deeply inspired by Whitehead's process philosophy. He widely applied the insights of the British philosopher to his models of organic development – defined as a spatio-temporal process-event. However, another fundamental figure who fostered Woodger's interests in the philosophical issues of biology was Hans Leo Przibram (1874–1944), during a term leave in 1926 in Vienna. Przibram had been a student of B. Hatschek (1854–1941), who in turn had been a pupil of R. Leuckart. After his graduation from the University of Vienna in 1899 and during his years of PhD study, Przibram bought the former Vienna Vivarium and converted it into a modern experimental laboratory – the Research Institute of Experimental Biology.[67] A trained morphologist, Przibram started to work on animal regeneration and transplantation after his experiences in Naples and Roscoff. After 1903, with the help of Paul Kammerer (1880–1926) as assistant,[68] he organized his experimental laboratory as an international centre where students and scientists such as Woodger could spend time doing their research.[69] This was a centre where the Kantian teachings were not forgotten because, as D. R. Cohen stresses, 'Diverse as the Vivarium's scientists were in their experimental approaches and in their philosophical convictions, they converged in their search for a "third way" between mechanical determinism and pure spontaneity, a framework that would do justice to the complex interactions between organism and environment'.[70] This was neither mechanism nor vitalism, but a synthesis allowing a scientific study of life without reducing organisms to chemistry and physics. This was the environment in which Woodger worked, and as J. R. Gregg and F. T. C. Harris explain, during his time in Vienna Woodger realized:

> that there were fundamental unanalysed assumptions in the theories then in circulation amongst biologists and that his training had not equipped him, nor was likely to have equipped anyone else, to examine or identify these assumptions. On his return

> to England he threw himself into a study of philosophy, or those aspects of philosophy that were the necessary pre-requisite for an analysis of biological theory. Within two years he had completed the necessary spadework and had gone on to analyse the assumptions implicit in the biological antithesis between vitalism and mechanism, structure and function, preformation and epigenesis, for example, and the theory of explanation in biology.[71]

Of course, all the figures I mention belonged to different generations and followed different trainings. However, as I shall show in the next chapters, they all shared a very similar conception of the organism and how the study of life should be carried out. They all had inherited, through diverse routes and ways, Kant's Romanticized bio-philosophical tradition. But they dealt with it in novel ways; they transformed and reshaped it, adapting these old teachings to new contemporary needs, knowledge and discoveries. In fact, we should not expect that twentieth-century biologists could use the same categories of thought or share the same preoccupation of nineteenth-century *Naturphilosophen*. Even though some of the early twentieth-century biologists still believed in the biogenetic law (or some less radical forms of it), accepted radical versions of holism and anti-mechanism, and maintained the importance of typological approach, they were not interested in the wild speculations that had characterized the biocentric philosophies of Schelling, Oken, Carus or Goethe, i.e. the primordial plant or animal, the absolute, or the sophisticated epistemology of Goethe and Schelling. Such an old tradition required a 'restyling' because problems and topics characterizing the old frame were changed or restated under a different perceptive. For instance, if Romantic natural philosophers had been deeply concerned with the problem of how a single cell can produce a complex multicellular whole, the new generation felt it more important to address the problems of heredity.

## Romantic Biology in the New World

In the United States, Germany was certainly one of the most sought-after countries of the old continent for its celebrated figures, bold *Weltanschauungen* and widely recognized scientific institutions. From the very first decades of the nineteenth century, German philosophy, culture and science began to be widely known in the New World. 'By the 1830s', B. Kuklick recounts, 'Germans ideas were making inroads at the American colleges. Many German immigrating to the United States had read German philosophy directly. Additionally, a trickle of American students returned from Germany with Kantian theories, and translations of German work were periodically published'.[72] Many of the most influential US philosophers had direct or indirect connections with German idealism. J. Dewey was deeply inspired by Hegel's philosophy, C. S. Peirce, as we have seen, by Schelling. Even W. James, who apparently disliked German thought, took inspiration from the philosophical speculation of Josiah Royce (1855–1916).[73] Royce was an American philosopher who had studied in Germany with the neo-Kantian philosopher Hermann Lotze (1817–81). Lotze had studied medicine and philosophy in Leipzig. He disseminated an original form of organicist philosophy in Germany and elsewhere, influencing, among others, T. H. Green and F. H. Bradley in England, and the well-known American philosopher G. Santayana.[74] However, the diverse forms of German-American idealism that emerged after the 1860s in the United States were frequently mixed with Darwin's and Spencer's evolutionism. Evolutionary theories often served as a blueprint for ambitious philosophical systems, worldviews that often inspired and appealed, in turn, to young zoologists. Static categories of thought, mind, knowledge, customs and morality – all could have evolved through adaptation, acquired characters and/or natural selection. As Dewey proclaimed in 1910: 'Doubtless the greatest dissolvent in contemporary thought of old questions, the greatest precipitant of new methods, new intentions, new problems, is the one effected by the scientific revolution that found its climax in the *Origin of Species*'.[75] Through evolutionary conjectures, biology entered deeply into philosophical discussions. Philosophers read biologists, and the latter were not immune from philosophical speculation. German idealism and British evolutionary ideas formed a powerful set of ideas that, in the hands of influential US thinkers, became grandiose schemes of an evolving purposive universe, life on earth, and the emergence of self-consciousness. Santayana's monumental *The Life of Reason: The Phases of Human Progress* (1905–6) is an explicit example: a bold synthesis of Hegelianism and evolutionism contained in five volumes.

It is therefore no wonder that, as a general rule, promising American zoologists in the last decades of the nineteenth century sought to spend some part of their overall training in Europe. They wished to go to France and England, of course, but most of all to Germany or German institutions. Germany was the country where many of the most exciting, advanced and innovative scientific discoveries were achieved. It was the place where new research directions in biology were identified and pursued, and where new experimental methods and practices were established.[76] Germany was the country from which came some of the most controversial theories of the organism: its development, its functions, its morphology and its inheritance. Germany was the country where Darwinism and Lamarckism had easily triumphed, thanks to Haeckel's tireless campaign and the wide diffusion of his successful bestsellers. Evolutionary morphology thrived, and the *Entwicklungmechanik*'s programme, first set by His and Roux in the second half of the nineteenth century, took form.[77] Finally, the mosaic theory of development advanced by Roux and Weismann, and Hans Driesch's hypothesis according to which the egg was a 'harmonious equipotential system',[78] renewed the old debate between preformationists and epigenesists. In brief, at the end of the nineteenth century, German biology was full of debates, contrasting research programmes, diverse traditions and innovative ideas, all elements that made this country very attractive for young biologists trained elsewhere. Many leading American universities held very good relationships with German institutions. The American students had the opportunity as well as the encouragement to spend part of their development abroad, gaining PhDs or doing particular research in Germany or in German institutions elsewhere, such as the celebrated Naples Zoological Station in Italy.[79]

However, among all of the widely recognized academic names and institutions, for some young promising American zoologists active in the second half of the nineteenth century, one figure was clearly their hero in zoology. This was not Roux, His, Haeckel, Gegenbaur or Weismann, but a figure less familiar to us: the previously mentioned German zoologist Karl Georg Friedrich Rudolf Leuckart (1822–98). Leuckart was a central figure in the academic German world – Nyhart ranks him as 'The most important morphologist to work as a professor of zoology', and indeed, 'by the early 1860s, he had become one of zoology's leading scientific lights; anyone who wanted a good introduction to state-of-the-art zoology went to Giessen to study with him'.[80] And yet:

> Leuckart exerted a profound and yet underappreciated influence on zoology as it diversified into new sub-disciplines. Researches from around the world came to work to his lab. Together with his collaborators and more than 115 doctoral students he mentored in his long career, they assured that his research approach and his ideas continued to flourish well into the next century. All in all, Leuckart's topics, interests, and functionalist approach to nature may be seen sprinkled through the works of his many students and co-workers.[81]

Leuckart earned his degree in 1845 at Gottingen University, working with the famous physiologist Rudolf Wagner.[82] In 1850 he was appointed associate professor at the University of Giessen, and in 1869 he moved to the University of Leipzig. While at Leipzig, in 1880 he was able to organize a new zoological institute and establish a research programme in which a sizeable community of international students was involved. This community eventually included figures such as the first future director of the Woods Hole Marine Biological Laboratory and University of Chicago professor Charles Otis Whitman (1842–1910), the future Harvard professor Edward Laurens Mark (1847–1946), the Harvard Medical School anatomist Charles Sedgwick Minot (1852–1914) and the Chicago physiologist Charles Manning Child (1869–1954).[83]

Leuckart was a deeply committed Kantian, and although a converted evolutionist, he remained loyal to the teachings of von Baer, with whom he remained on good terms until his death.[84] The attachment to von Baer is important because Leuckart – along with Kant, von Baer and other embryologists in general – held that the unity of the organism was comprehensible only through the study of its development. Thus as a prominent morphologist and, at the same time, physiologist and embryologist, Leuckart subscribed to all the organicist principles I have previously mentioned. To briefly recapitulate them: biomedical sciences had to be independent from the physical sciences. Organisms are not simple mechanisms insofar as they exhibit complex and irreducible forms of causal interaction. Although vital forces are not required, living organization must be considered as given. Yet, and as a consequence, the organic whole is more than its constituent parts, i.e. parts work in function of the whole so that an organic individual must be understood as a teleological system. Finally, the organism had to be seen in constant and dynamic interaction with its environment.

Leuckart first established his organicist synthesis in a monumental work written with the anatomist and physiologist Karl Bergmann and published in 1852: *Die anatomisch-physiologische Übersicht des Tierreichs* (An Anatomical and Physiological Overview on the Animal Kingdom). In this work the authors adopted a physiological approach to the study of the animal form.[85] Indeed, drawing directly on the works of the philosopher Hermann Lotze and the chemist Justus von Liebig, as well as on Kantian teleology, they forged a synthesis in which mechanical and teleological explanations in the life sciences could find a profitable relationship. Organisms had to be conceived as material entities, although their understanding required, at the end of their investigation, a teleological framework. The functional morphology Leuckart set out was quite successful. Ever since the publication of his *Ueber die Morphologie und die Verwandtschaftsverhältnisse der wirbellosen Thiere* (On the Morphology and Relationships of the Invertebrates) in 1848, he had defined morphology as the science that would find a middle way between the embryological approach of von Baer and the teleological and functional approach of Cuvier.[86]

Although interested in theoretical and philosophical issues surrounding life science, Leuckart spent the last years of his career studying parasitic worms. He is indeed nowadays mostly celebrated as a parasitologist rather than a zoologist or physiologist. However, his intellectual legacy cannot be overlooked, especially regarding his influence on the students he trained. As Nyhart rightly reports, 'Leuckart continued to defend the language of purpose as useful heuristic, and his students did the same';[87] in fact, as we will see in the next sections and chapters, some of his American students remembered Leuckart's lessons well. Indeed, we can easily track the conceptual chain through which Leuckart's biology was directly or indirectly diffused in the United States. Firstly, Whitman went to Leipzig in 1875 to study for his PhD with Leuckart.[88] After a brief experience in Japan, Whitman returned home and became one of the most relevant defenders of organismic biology in the United States (as we shall see later on). In 1892 he was appointed professor at the Zoological Museum of the University of Chicago, and there he mentored F. R. Lillie and hired C. M. Child.

Frank Lillie became a staunch collaborator of Whitman at the University of Chicago and at the newly founded Woods Hole. Starting as assistant director there, from 1908 Lillie replaced Whitman as director at Woods Hole. As I will show in Chapter 5, Lillie acquired Leuckart's lessons through Whitman's important influence. Child instead went directly to Leipzig in 1892. In 1894, after concluding his PhD, Child spent a semester at the Naples Zoological Station (another German institution), and in 1895 he was hired by Whitman as a zoological assistant at the Zoological Museum of Chicago. Hence not only had Child directly absorbed Leuckart's teaching in Leipzig, he also was in contact with someone who had studied with Leuckart and disseminated his agenda in the US. The zoologist E. L. Mark (1847–1946) had also been a pupil of Leuckart in Leipzig, and in 1877 he was hired by Agassiz at Harvard's Museum of Natural History. As Winsor remarks, 'Mark, with his new PhD from the University of Leipzig under his arm, was determined to bring back home the latest techniques in cytology and the high standards he had learned in Rudolf Leuckart's laboratory'.[89] W. E. Ritter (1856–1944), a central figure in my story (see Chapter 6), was a pupil of Mark at Harvard. After concluding his PhD in 1894, Ritter sailed to Europe, where he worked at the Naples Zoological Station and then at Berlin University. In Naples Ritter befriended Child, with whom he remained in contact for the rest of his life. In short, Ritter, as did the all other figures I mention, worked with someone who was directly acquainted with Leuckart's biology.

Ernest Everett Just was a younger American organicist who I will also consider in this book. He was awarded his PhD at the University of Chicago in 1916 under Lillie. During his stay in Chicago, he also attended Child's lectures in physiology. However, in 1928 Just went to Europe to work at the Naples Zoological Station, at the Kaiser-Wilhelm-Institut für Biologie in Berlin (today the Max Planck Institute, relocated to Tübingen in 1945) and finally, at the French Marine Laboratory in Roscoff (a place where Delage, a developmental and famous organicist biologist, had worked for a long time). Just's experience in Europe was a determining factor in his bio-philosophical formation, as K. R. Manning's excellent biography and Just's 1939 book attest (see Chapter 5). Thus, thanks to Leuckart and his international laboratory in Leipzig, the Romantic bio-philosophical tradition thrived in the US. It was never forgotten, as Lenoir argues, but rather transformed and adapted to another context. This American group of neo-Romantic scientists, people such as Child, Ritter, Lillie and Just, aimed for an organismic version of biology characterized by a strong anti-reductionist approach, an approach that clashed with the newly emerging theories of heredity, evolution and development.

## New Wine in Old Bottles

In this chapter I have shown that Kant's bio-philosophy neither died with the German Romantic naturalists and philosophers, nor extinguished with Leuckart in Leipzig, but instead that it thrived through different routes and ways both in the UK and the US. Of course, Kant's tradition as understood or accepted by Müller and Leuckart in the nineteenth century was not the same as that of Ritter or Russell, Just or Woodger in the twentieth. New challenges, discoveries and issues would necessarily impact upon such an old theoretical framework. However, the general questions remained; the organicist principles I mentioned before continued to be discussed and considered controversial themes. Although after the *Naturphilosophen* and the Romantics, Kant's strategy of reason was transformed into an ontological stance; although the issue of heredity acquired an increasing importance; and although the relation of parts to the whole was then reframed as relations in a complex system-hierarchy, the way of thinking about biological problems remained profoundly Kantian. This characterized part of what in the twentieth century Morton Beckner has dubbed 'The Biological Way of Thought'.[90]

In the next chapters I will consider some twentieth-century figures who shared that way of thinking. However, even though I define this group of thinkers as 'Romantic biologists', with the expression 'Romantic bio-philosophical tradition' I refer to neither a well-defined generation of scientists nor a circle of individuals sharing the same training, coming from the same universities or laboratories, or heirs of one unique school of thought. Instead, with such a label I embrace all those individuals discussing the mentioned organicist approach, sharing a similar theoretical universe, engaging with each other's work, and proposing alternative views on heredity, development and evolution - alternatives rooted in their organismic convictions. In fact, I am entirely conscious that between Russell and Woodger or between Child and Just there was a quite extended generational gap. However, I will show that although they belonged to different academic worlds and contexts, they all shared methods, approaches to science and, especially, a way of thinking about particular issues. To them, for example, as I emphasized in the previous chapter, biology had to be thought of as an independent science with its own rules, goals and objects. Yet they not only shared most of the organicist principles previously mentioned, they also understood the notion of heredity in a very particular way. In fact, for all of these 'neo-Romantics', heredity – to use an analogy – was the term replacing Wolff's *vis essentialis*, Blumenbach's *Bildungstrieb*, von Baer's *Trieb* or Müller's vital force, or any un-analysable and irreducible cause supposed to lie behind a living organisation. In other words, heredity came to be represented as the material *quid* explaining the origin of a living organisation, as well as its persistence and its continuity through generations. It is for precisely this reason that, to them, the central question to pose for any science of heredity based on organismic principles was not about the ratio-distribution of visible characteristics transmitted from generation to generation; it was the fact that individuals persist and insist on forming the same complex functional structures. As Just clearly and ironically said, the mystery of heredity was not solved by counting the bristles on *Drosophila*'s back, or by the transmission of its eye pigment, but by understanding how the embryo forms its back and its eyes.

Twentieth-century organicist discussions owed most of their philosophical sophistication to the Hegels and Schellings of the twentieth century: the French philosopher and Nobel laureate H. Bergson and the British mathematician A. N. Whitehead. Bergson's creative evolution and Whitehead's philosophy of the organism provided the intellectual scaffolding for new scientific syntheses in the twentieth-century life sciences. With Schelling, Bergson saw life as pure creativity. He saw evolution as process of spontaneous productivity leading to a superior form of freedom from a mechanical and determinist universe. Life was conceived as pure change, tendency and struggle to overcome the inorganic world. Life defied any fixed category of thought and could be gasped with neither mechanistic hypotheses nor grandiose teleological narratives. Indeed, Bergson rejected both mechanism and universal teleology. The two stances were accused of considering evolution as predetermined since the beginning. Instead, the evolution of life was a contingent narrative. Life was characterized from an irreducibly creative and spontaneous *élan*; it created new forms through an irreversible and unpredictable process.[91] With Kant and many nineteenth-century Romantic naturalists, Bergson believed that embryological phenomena demonstrated the irreducible purposiveness and unity of organisms, a purposiveness and unity that could neither be reduced to physics and chemistry nor grasped with the help of neo-Darwinian speculations. Life, Bergson insisted, has a tendency to individuality; it tends to build close and dynamic systems, systems functionally able to adapt themselves to the changing conditions of the environment. As a consequence, he believed that the only hypothesis able to explain the inner purposiveness of the organism was a *psychobiological* neo-Lamarckism (see Chapter 4). Organic systems change because they have an inner capacity to adapt themselves, to modify and transform their integrated structures and functions to the constant alterations of the external environment.

As is well known, Bergson's *Creative Evolution* (first published in 1907) was a sensation.[92] In England, the philosophers H. W. Carr and R. G Collingwood and the zoologist Thomson widely disseminated Bergson's philosophy. Haldane, D'Arcy Thompson, Russell and many other biologists were deeply inspired by Bergson's sophisticated form of vitalism. However, Whitehead's organic philosophy was no less influential. From the 1920s onwards, Whitehead began to develop a metaphysical doctrine that had to replace materialism and mechanism. The next stage for scientific realism, Whitehead thought, had to be based on the concept of 'organism' because the cosmos resembles less a mechanism than an organism: 'nature is a process of expansive development ... nature is a structure of evolving processes'.[93] Romantic naturalists, poets and philosophers, Whitehead noted, had already criticized the mechanist and materialist views of the eighteenth century. They had anticipated a form of organic philosophy that had to replace a valueless and mechanic world with a creative and dynamic universe. With Bergson, Whitehead believed that evolutionary philosophy was inconsistent with materialism. Mechanists and materialists assumed a substantialist metaphysics that clashed with the intrinsic creativity of evolution: 'Evolution, on the materialistic theory, is reduced to the role of being another word for the description of the changes of the external relations between portions of matter ... there can merely be change, purposeless and unprogressive'.[94] Whitehead's organic metaphysics attracted many biologists. Haldane, D'Arcy Thompson, Russell, Woodger, Ritter and many others all used the ideas and insights of the British philosopher.

But organismal biologists were also deeply engaged in discussions about the causes of speciation in evolution. Between the 1930s and 1950s a new synthetic theory of evolution emerged, an ambitious theory that J. Huxley dubbed 'modern synthesis'.[95] One of the main points characterizing the movement was that speciation was explainable as stochastic variation and selection of genes in a population over time. Individual development was of no use for the understanding of evolutionary diversification. Development was all that followed from the genetic information, and the proximate cause of form, to use Mayr's terminology.[96] But for many figures who belonged to the tradition I am describing, all that made no sense. Evolution had little or nothing to do with the stochastic distribution of particles called genes. Instead, the emergence of evolutionary novelties was conceived as a process related to individual development. Variations did not arise from gene mutations but from embryogenetic disruptions, deviations, divergences and systemic alterations that could be transmitted to the next generations. Most of the architects of the modern synthesis placed populations – with all their intrinsic variations – at the centre of their evolutionary discourse; populations, not individuals, were the main ontological element that all evolutionary biologists had to assume. However, for the Romantic biologists, the principal ontological category to assume was the individual organism, with its complex and dynamic development; its capacities to adapt itself to very diverse environments; its irreducible organization; and its creative behaviours. As a consequence, evolution consisted of the progressive and systemic changes of developmental trajectories. For them, nothing in biology made sense if not in the light of development (including evolution).

# 3 THE BRITISH VERSION: J. S. HALDANE, D'ARCY THOMPSON AND THE ORGANISM AS A WHOLE

In the previous chapter we saw some of the possible routes through which the organicist tradition travelled. I sketched, in a very general and synthetic way, the network that from Kant onwards guaranteed the diffusion and extension of a specific research programme that – notwithstanding its idiosyncrasies and contradictions – directed the aims and methods of many naturalists active in the nineteenth and twentieth centuries. In this chapter, we will see more in detail the way this tradition (which I have dubbed Romantic biology) was translated, transformed and re-proposed by two important scientists mainly active during the first half of the twentieth century: J. S. Haldane and D'Arcy Thompson. As I mentioned earlier, when Haldane delivered his paper at the History of Science Congress in 1931, he was recognized as one of the leading physiologists in England. Born in Edinburgh in 1860, Haldane studied medicine at the University of Edinburgh and, like many biologists of his generation, spent part of his training in Germany, particularly in Jena, where he attended the lectures of J. W. Dobereiner, J. Loder, E. Haeckel and the famous botanist E. Strasburger.[1] Throughout his life, Haldane was a staunch supporter of German biology and philosophy. He loved and praised Goethe's philosophy and science, and acquired a deep knowledge of Kant's critical philosophy and Hegel's idealism through the acquaintance of the British idealists.[2] In his first publication, an article written together with his brother Robert, Kant's bio-philosophy was explicitly discussed.[3] One of the venerable conclusions of this early work is striking and demonstrates the overwhelming pervasiveness of Kantian philosophy and its wide diffusion:

> For Kant then, the relation of reciprocity, as the most concrete of the categories, is the highest relation of reality ... The fact is that every part of the organism must be conceived as actually or potentially acting on and being acted on by the other parts of the environment, so as to form with them a self-conserving system. There is nothing short of this implied in saying that the parts of the organism can adapt themselves to one another and to the surroundings.[4]

Philosophical speculations would always be an essential part of Haldane's scientific career, even in his most technical and specialist books and papers. Indeed, as we will see later, the organismic approach he supported informed all of his physiological results and experiments.

In 1884, invited by Thompson, Haldane moved to the University of Dundee. They had been close friends from a young age, when they founded a naturalist club called 'Eureka', a club including among its twelve members not only Haldane and Thompson but also William Abbot Herdman (1858–1924), who would become a reputed marine zoologist and oceanographer at the University of Liverpool.[5] As Thompson's daughter recalls, they used to spend every Saturday together, 'botanising through the countryside, howking fossils in quarry and railway-cutting, grubbing in the rock-pools at Wardie, or searching the jetsam of Newhaven fishing boats'.[6] When Haldane arrived in Dundee, Thompson held a chair as Professor of Biology there and was actively involved in research regarding hygienic and material conditions affecting the local population. Together with Haldane, Thompson organized a small laboratory to study the pathological living conditions in Dundee's slums; then Haldane and Thompson, with a policeman as guide, wandered in the night through local slums, gathering information and experimental data in order to assess the quality of air breathed in the overcrowded and undersized houses occupied by poor people.[7] These experiences shaped Haldane's future interest and expertise in the physiology of respiration as well as his lifelong concern for working-class conditions. Haldane tried to always match his philosophical and theoretical interests with concrete and practical science – a science aimed at improving the life and working conditions of the most disadvantaged people.

Even though Haldane moved to Oxford shortly afterwards – after spending a few months at the University of Berlin's Pathological Institute, attending the lectures of Bois-Raymond – his friendship with Thompson continued. In fact, although interested in very different topics and having diverse opinions about life sciences and methods of investigation, they shared common views on what biology should be. Thompson had been a pupil of Balfour in Cambridge and had had an embryological formation. During his Cambridge years, he had become a friend of Whitehead and Sherrington,[8] and once in Dundee, he worked together with Geddes. Yet like Haldane, he was a deep admirer of Henry Bergson's and General Jan Smuts's work,[9] and also like Haldane, he loved German biology[10] as well as philosophical speculations. For example, in 1884 Thompson published a short article in *Mind*, discussing Haldane's philosophical interpretation of animal regenerative phenomena.[11] Thompson observed that Haldane:

> makes use of the phenomena of regeneration of lost parts in contending that we must sublate our view of the nature of life from a mechanical or cause-and-effect category to a category of reciprocity, and from that further to one in which the parts must

> be regarded as determined in relation to an idea of the whole; in other words that organic processes cannot be reduced to series of causes and effects, or to what is the same thing, matter acting as a vehicle of energy.[12]

After a list of regenerative phenomena (beginning with Trembley's *Hydra*), Thompson concluded that in the process of animal regeneration, even though the cells seemed to follows a specific 'custom' or 'habit' acquired during the organism's phylogeny, and even though some mechanistic explanations were possible, Haldane was right in arguing that the whole played an essential role in shaping the parts during regeneration.[13] Even though some hereditary tendencies could mechanically explain the re-formation of lost parts (as Weismann had argued), there were phenomena where the influence of the whole organism was required to explain particular cellular arrangements and behaviours.

Many years later, in 1918, both friends participated in a symposium held at the Aristotelian Society in London. The theme was very philosophical: 'Are Physical, Biological and Psychological Categories Irreducible?' The debate was introduced by Haldane himself, who quickly dismissed the hypothesis according to which biological categories could be reduced to physical ones; indeed, in echoing Kant's reflective judgment, he argued:

> it is only through the central working hypothesis or category of life that we can bring unity and intelligibility into the group of phenomena with which biology deals; and it is because the biological working hypothesis is for the present absent in our ordinary conceptions of physical and chemical phenomena that we must treat physical and biological categories as radically different.[14]

For Haldane, just as biological categories (or working hypotheses, as he called them) could not be reduced to physical categories, so psychological categories could never be reduced to biological ones: all three sciences required different theoretical frameworks insofar as they dealt with different phenomena. However, complex organisms included all three levels insofar as any physiological reaction of a conscious organism, within its environment, was not solely determined by physical and biological objects; there were also subjective elements affecting possible response-reactions.

In Thompson's reply, there were some apparent points of disagreement. Firstly, Thompson carefully reported and synthesized Haldane's position; then he declared that psychology should be kept outside the discussion because mind and matter are incommensurable.[15] However, things were different concerning physics and biology. Indeed, Thompson argued, there was not a necessary incommensurability between the two; the physicist's concepts and methods were of fundamental importance for biologists. Organisms, after all, were plainly part of the physical world and subject to physical and material forces: 'I know that change and form in a concrete material body involves the movements of matter, and the movements of matter are to be symbolically ascribed to the action of force, and actually to the transference of energy'.[16] Thompson did not support physical reductionism; what he was arguing was that even though organisms were mechanisms and could be described or understood in mechanical terms, they were nonetheless mechanisms of a very special kind.[17] Biological categories could include mechanical analogies, but not in the ordinary sense that, for instance, Cartesian tradition intended. In fact, the meaning of 'mechanism' itself was in constant change.[18] Haldane replied to Thompson and totally agreed with him: 'I fully accept Professor Thompson's suggestion that it is only the "ordinary" working hypothesis of physics and chemistry that seem to me inadequate in biology. Recent developments of experimental physics and chemistry are profoundly changing these conceptions, and, as it seems to me, tending to bring physics and chemistry not only much closer to one another, but also much closer to biology. Here, then, our differences seem to disappear'.[19]

The discussion between Haldane and Thompson about the nature of life phenomena continued. Thompson read and commented on Haldane's work, eventually writing critical and accurate reviews of his books. For example, in 1919 Thompson wrote a review in the journal *Mind* about the controversial new work Haldane had just published that year: *The New Physiology*. Thompson acknowledged the radical potential of the book: 'Dr. Haldane is in revolt with much more than the tenets and the methods of the modern biology; it is a larger philosophy that he has mind, and his challenge is to the world'.[20] Although he accepted Haldane's teleological approach to medicine and biology, and his practical recommendations to physiological practices (in fact, in medicine, as much as in all phenomena of life, the Aristotelian *Telos* – the final cause – seems sometimes to be an inescapable principle to which any investigator must succumb), Thompson felt that Haldane's position was too strong. The rejection of all physico-chemical approaches purely because they could not give a comprehensive understanding of all living phenomena betrayed Haldane's impatience, Thompson complained. Even though Thompson felt himself heretic regarding some biological theories, such as cell-theory, he maintained that the quest for mechanical causation still remained open, even though it could not be universally applicable.[21]

The discussion on mechanism and other possible alternatives in biology continued, in England and elsewhere. Ten years later, in 1929, both men, now old, participated at the British Association conference in South Africa. Haldane participated as lead speaker in a session titled 'The Nature of Life'.[22] Haldane and Thompson, together with other speakers and many participants, discussed the subject with the notorious Cape Town host, Jan Christiaan Smuts, who three years later published a rather controversial and provocative book entitled *Holism and Evolution*. The meeting was in fact a joint session of physiology, botany and zoology,[23] and Smuts opened the debate with his new, holist hypothesis of life and evolution as opposed to mechanist approaches in biology.[24] L. Hogben played the devil's advocate, strongly arguing for reductionism and mechanism in biology.[25] Unfortunately we have no clue about Thompson's contributions to the South African session; we only know that organismal or holist interpretations of living phenomena were always at the centre of his interests.

## A Romantic Physiologist

In the preface of his *Philosophy of a Biologist* (1935), Haldane explicitly acknowledged his intellectual debts: 'my special obligations to British contributors to post-Kantian and post-Hegelian thought will be evident, and particularly perhaps T. H. Green and F. H. Bradley'.[26] In fact, throughout his life Haldane had always seen himself as a pupil of Kant and a well-acquainted interpreter of the British idealist tradition.[27] Haldane returned to study in Germany several times, and he had an enthusiastic and expert Germanophile in his own family. His older brother, Richard Haldane, had studied in Gottingen with Hermann Lotze and, years later, would publish the first British translation of a book by one of the most controversial of Kant's scholars: Schopenhauer's *World as Will and Representation.* Lotze was a key figure for both brothers. Drawing on Kant, the German philosopher had proposed an original synthesis between mechanism and teleology. He thought that the two implied one another. As Lotze remarked: 'Mechanism has therefore no power controlling reality, but rather the way in which a purposeful reality realizes itself'.[28] It would be no exaggeration to say that the young Haldane 'breathed' German science and philosophy within his family and the intellectual context in which he lived.

One year after he had published the article with his brother, Haldane, then twenty-four years old, published a new paper in which he denounced the general tendency of physiology to reduce all organic processes to a well-defined series of causes and effects.[29] Such an approach, Haldane argued, rested on the false analogy of the organism seen as a machine like, for example, a steam engine. However, even the simplest organism, such as the common earthworm, showed 'purposive' behaviour which could never be reduced to a machine's behaviour; Darwin, Haldane continued, had clearly showed that earthworms act in intelligent and unexpected ways because they are able to adapt their actions to unusual circumstances. Not only do earthworms exhibit purpose, but the regeneration of tissues and limbs in many animals, especially lower forms, also demonstrates an elastic and purposive action of the whole organism on its parts. To the young Haldane, the behaviour of the organism and other phenomena related to reproduction, development and regeneration showed that the category of cause and effect, as an instance of a misplaced mechanical analogy, could never be applied to the study of life phenomena, because organisms had to be seen as systems in which the parts 'are determined, not only as regards their reciprocal action on one another, but, also as regards what is inherent in the part themselves of a system whose parts reciprocally determine one another'.[30] The right category for studying organisms was that of 'reciprocity', in which the organism was built not through a web of consequential causes and effects but though a system of 'reciprocal determinations'. Haldane concluded with an example taken from one of his favourite disciplines: physiology. In fact, the well-known physiological phenomenon of reflex action,[31] and the hypertrophy of muscular organs (and other kinds) when corresponding organs are disrupted by injury or inflammation, all show that a mechanistic approach fails in explaining these phenomena, and physiologists, like biologists, require an 'organismic' approach to life phenomena.

In the following years, Haldane's programme became more and more convincing, broad and original; from a well-circumscribed critique of mechanist and reductionist biology, he began to conceive a broad 'organismic' agenda which included not only biology and physiology but also medicine, psychology and even politics. In 1913, when Haldane was a well-established international physiologist working in Oxford, he published one of his first great syntheses: *Mechanism, Life and Personality*. The book was a result of four lectures delivered at the Physiological Laboratory of Guy's Hospital in London in May of the same year. In these lectures Haldane developed some of the tenets characterizing the organicist bio-philosophical tradition: denial of vitalism and mechanism; the acceptance of living organization as a postulate; the use of teleological explanations; the essential and dynamic relationship between parts and whole; and the fundamental interrelation between organism and environment. In particular, regarding physiology, Haldane drew on J. Müller and his 'famous' old textbook, as he considered Müller to be one of the great heroes of physiology and developmental biology.[32] Haldane also introduced one of his most recurrent themes characterizing his holist view of life: physiology of respiration. Respiration demonstrated the intricacy and interconnection of many different factors involved at the same time (I will discuss this in the next section).

Once he had established his general organismic framework in physiology or other disciplines with which he felt more acquainted, Haldane criticized Weismann's theory of heredity and development, which he considered 'little more than empty words'.[33] Finally, he concluded with a fairly long discussion of Kant and his transcendental theory of experience, and began framing his own, original idealist philosophy of the individual and personality. He argued, among other things, that our world of experience is first and foremost a world of personality, namely, a world where man is considered primarily as a person 'in conscious relation through perception and volition to his environment'.[34] For Haldane, in effect, both physical and biological realities were mere abstractions of the world of personality, a world scientifically open to the psychologist's investigations. This philosophical scheme allowed Haldane to justify and reinforce the presumed independence of life sciences from physical sciences. Furthermore, it allowed him to open the way for a general discussion about epistemology in life sciences, namely, the way we perceive life and gain knowledge of it.

## Haldane's 'Concrete' Physiology

> Those who in their innocence cling, on theoretical grounds, to a mechanistic conception of the passage inwards of oxygen are straining at a gnat while swallowing a camel.[35]
>
> J. S. Haldane

Although, in his later years, Haldane continued to be interested in philosophical problems surrounding science and society, he had a very concrete vision of how theoretical concerns dealt with practical issues. His science, far from being a quest for disinterested and abstract topics, was a search for applied knowledge: he invented technological devices like gas masks; he introduced safety measures and apparatus for miners, inventing the famous 'Haldane box';[36] and he made fundamental contributions to diving techniques, formulating the diving decompression tables.[37] All his contributions to science and technology were mainly related to his studies on the physiology of respiration, and this became the prototype of his organismic physiology that, in turn, brought him to criticize Weismannian and Mendelian sciences of heredity as well as British individualist society. Indeed, between 1917 and 1921 he published three important books, all mainly dealing with the physiology of respiration: *Organism and Environment as Illustrated by Breathing* (1917), *The New Physiology* (1919) and *Respiration* (1921). By the time the first of these books was published, he had acquired an outstanding knowledge out in the field: he had worked in the sewers of Dundee and London, studying the so-called 'sewer gas'; he had explored working conditions in the deep caves of coalmines (and tried to develop measures to avoid mine disasters); he had performed expeditions to the highest available peaks in order to study the processes of physiological acclimatization (in particular, Ben Nevis in Scotland and Pikes Peak in Colorado in 1911); he had directly experienced trench warfare in World War I in order to find measures against German poisonous gases; he had worked and was a member of the 'Deep Sea Diving Committee' in order to improve deep-sea diving techniques; and, finally, he had made experiments on himself in his lab in Oxford. As Goodman remarked, Haldane brought his lab into the field and, I would add, applied his theoretical conceptions to concrete inventions. Through all these experiences, Haldane was even more convinced that breathing was essentially a phenomenon of organic self-regulation, entailing many different interrelated elements.

Throughout the nineteenth century, as G. E. Allen observes, there were two main hypotheses about respiration: on the one hand, it was thought that breathing depended principally on the activity of the nervous system, which worked as the main stimulus for respiration. On the other, it was assumed that respiration was stimulated by some specific chemical substances dissolved into the blood through tissues.[38] Both hypotheses presumed that breathing represented a delimited or circumscribed phenomenon caused by a well-defined underlying mechanism. Haldane totally challenged such an assumption; in the phenomenon of respiration, many other causal elements were involved (including metabolism). Indeed, even though the mere act of breathing could be seen as a series of muscular movements of the chest following a well-defined rhythm, and even though the rhythmic exchanges of oxygen and carbon dioxide could be understood in physico-chemical terms, respiration seen as a whole process implied the ordered functioning of other organs, tissues and bodily fluids such as blood.

Oxygen, once absorbed in the lungs, passes into the blood and is carried and distributed throughout the body by capillary vessels. Afterwards, the unoxygenated blood is carried back through the veins to the heart and lungs in order to acquire new oxygen and repeat the cycle. However, muscular rhythmic movements related to respiration are regulated by nerve stimuli produced in a central area within the medulla oblongata; such an area is in turn influenced by chemical states of the blood coming from the heart to that central area. Chemical states depend on the rate of carbon dioxide previously dissolved in the blood, which causes a weak alkalinity of arterial blood so that the central area reacts to small differences of alkalinity. The weak alkalinity in the blood depends, in turn, on the presence of other substances in the blood, for instance the presence of salts, which is regulated by a balance between absorption in the intestines and excretion by the kidneys. The kidneys therefore have a fundamental importance in regulating respiration because they control the rate of some essential substances which, in turn, influence the central area in the medulla oblongata. Of course, even though salt excretion by the kidneys depends on alimentation and other external factors, the organism is able to regulate its internal environment despite changes in the external environment; therefore, Haldane added, internal organic balance is not necessarily due to external constancy of the environment. The result, as Haldane argued, is that we play 'an aimless game of battledore and shuttlecock'.[39] Respiration, like many other related organic processes, shows that the 'phenomena represented by a living organism cannot possibly be grasped and interpreted one by one ... We must grasp them as a whole, simply because, whether we will or not, they present themselves to us as an organically determined whole'.[40] To Haldane, respiration did not represent a discrete physiological process among others; it represented an instance of the general organic self-maintenance processes of the organism as a whole.

The holist interpretation of physiology was applied to as well as reinforced by Haldane's direct experiences and observations on the field, from the extraordinary abilities of miners to work in unimaginable conditions of stress and heat, to divers' remarkable powers of adaptation to different pressures and ever-changing circumstances of a hostile environment. When, in 1911, Haldane reached the peak of Colorado's pikes, he could experience, in his own body, the capacity of the organism of self-adjust and self-regulate as a whole to the decreasing amount of oxygen in the air.[41] The well-known phenomenon of acclimatization, which he had discussed extensively in his *Organism and Environment*, proved that many living phenomena and abilities had a purposive tendency; in fact, in higher altitudes the lungs' epithelium tends to absorb more oxygen, and the rhythm of breathing and concentration of haemoglobin increase. Then 'the last raises the oxygen pressure in the capillaries of the body, the second diminishes the fall in alveolar oxygen pressure, and the first raises the arterial oxygen pressure much above the alveolar oxygen pressure, whereas at the sea level the arterial oxygen pressure is no higher, as a rule, than the alveolar oxygen pressure. *The teleological significance of these changes seems clear*'.[42] In short, a wide range of phenomena (acclimatization, breathing and other phenomena of physiological regulation), the extensive capacities of organisms to recover function and organic balance (as Haldane had directly observed in men recovering after gas intoxication or infected wounds), and the unexpected vicarious functions of organs and developmental phenomena all spoke in favour of organismic biology and physiology.[43]

## From Physiology to Medicine

> Had I been content with mechanistic interpretation I should not have suspected the presence of various defects in existing knowledge, nor known in what direction to look for them.[44]
>
> J. S. Haldane

Neither biologists nor physiologists who accepted the organismal approach to biology could fail to see the relevance of it for medicine. In fact, when Haldane read a paper at the Edinburgh Pathological Club in January 1918, he made clear that there was a very close connection between physiology and medicine. He first recalled the therapeutic measures he used for curing soldiers exposed to lung-irritant gases. These gases caused poor oxygenation in the affected lungs, which, if prolonged, damaged the nervous system, causing haemorrhages in the brain as well as injuring other vital organs. However, if oxygen were administrated, all the damages following want of oxygen could be easily avoided: 'What we do by giving oxygen is to keep the patient alive, by preventing secondary injury, until time has been given for the natural processes of recovery of the lung'.[45] After all,

Haldane continued, surgeons and physicians of different specialities worked in the same way and pursued the same scope: in treating wounds antiseptically or aseptically, they prevent infections and allow these injuries to heal themselves; the use of drugs or anaesthetics was intended to break a vicious and dangerous circle and permit the body to regain the balance lost in disease; and in administrating *Digitalis* and *Strophanthus*, physicians helped the diseased heart to regain its natural equilibrium. There are many other examples, but what Haldane was essentially arguing was that 'We cannot repair the living body as we repair a table or a clock, the surgeon is not a carpenter, nor the physician a mechanician or chemist';[46] indeed, he continued, 'in practical medicine the assumption of the existence of organic regulation of structure and function is absolutely fundamental. Disease is the breakdown of this regulation at one point or another; and practical medicine is simply assistance to Nature in restoring and maintaining effective organic regulation'.[47]

Haldane noticed that physicians did not consider physiology a central discipline because it was taught in physico-chemical terms, whereas practical medicine was clearly based on a teleological conception of the organism. However, in the very last years of his life, when he published an essay in 1923 denouncing, once again, the artificial gap between physiology and medical practice, he insisted that Institutes of Medicine had to be based on a new physiology: organismic physiology.[48] In fact, doctors, like biologists, always deal with an 'actively persistent relationship' which requires a deep physiological knowledge of what is considered 'normal regulation', namely, the normal, not pathological, functional state of the organism as a whole.[49] The new physiology, as conceived by Haldane, had to become the keystone of medical sciences because it was the only discipline giving the appropriate foundational knowledge for dealing with self-regulated systems; in fact, as Hippocrates had taught centuries before, curing means re-establishing a balance in a disrupted body.[50]

## Against Weismann and Mendel

The organicist view of life that Haldane professed not only supported his new conception of physiology in relation with medicine, as it had done before with biology, but it also guided his ideas about heredity and his critiques against Weismann's germ plasm and Mendelian genetics. In 1890 Haldane went in Fribourg and attended Weismann's lectures; however, as he recalled many years later in one of his Donnellan lectures delivered at the University of Dublin in 1930, he remained rather intellectually disappointed: 'Attracted by his vigour and genial personality, I went to work under Weismann when I was a young man; but I soon realised that, since it is actively maintained structure that we are dealing with, we cannot hope to interpret living structure, whether in the embryonic or adult stage, apart from its environment, although, since we are also dealing with actively maintained life, neither can we disregard heredity'.[51] Haldane regarded Weismann's theory of germ plasm as an instantiation of the mechanistic approach to biology, which was for him misleading and wrong. It was so because such an approach tended to separate the phenomena of life into well-defined compartments: heredity, as conceived by Weismann, was a phenomenon limited to the cell nucleus and its substances, elements that determined all of the complex changes happening during development. In other words, Weismann's germ plasm, with all its sets of supposed determinants, pretended to explain not only the organism's reproduction, but also living organization. However, as Haldane contended, such a view ran against the most basic conception of life as regarded from an organismic viewpoint. In fact, as we have seen, Haldane's organicism maintained that causes and effects, as well as stimuli and responses, all constituted an interrelated system forming a unique organized whole. Instead, as C. M. Child was arguing from Chicago, all phenomena of life (including heredity) were based on metabolic changes, and Haldane was of the same opinion:

> Biology deals at every point with phenomena which, when we examine them, can be resolved into metabolic phenomena – exchange of material and energy, as exemplified in growth, development, maintenance, secretion and absorption, respiration, gross movements in response to stimuli, and other excitatory processes ... There is no reason for separating the reproduction of a whole organism from the constant reproduction of parts of it in ordinary metabolic process. *Hence our conception of heredity involves every part of biology*.[52]

Because it was impossible to separate such a specific substance from the whole organization, and at the same time, insofar as it was impossible to imagine that such a substance was responsible for the organization of the whole, Weismann's theory was rejected. Haldane never changed his mind. In 1932 he returned to this topic in a paper delivered at the British Institute of Philosophical Studies; after harshly criticizing Weismann and his advocates, he advanced his own suggestions of how research on heredity should be carried out. The student of heredity should never try to identify one substance, one single stuff, one unique cause explaining transmission and development, because during reproduction all living activities are involved: 'What is reproduced is living structure and activity with all their delicacy of detail'.[53] All particulate theories of heredity, along with mechanistic approaches to biology, necessarily bring back vitalism and mysticism because, ultimately, these theories always assume what they intended to explain. Heredity, like respiration, was just one instance, one manifestation, one characteristic of a unique interconnected system; therefore heredity was a *systemic* phenomenon, not a transmission of material stuff.

Weismann was not alone on Haldane's blacklist. In a lecture delivered at the Birmingham and Midland Institute in 1917, he also attacked both Mendel and Bateson. First of all, stated Haldane, insofar as the organism and environment are one unique dynamic thing, environmental changes in adult organisms can be fixed and transmitted in the germ cells, so that 'acquired characters must be capable of hereditary transmission'.[54] Indeed, Weismann's hypothesis, according to which there was a physiological separation between germ and soma cells, left the origins of variation unexplained. In order to overcome this difficulty, Haldane argued, Mendelian heredity was invoked. Because Mendel had argued that characters or factors were transmitted and distributed generation after generation without being blended, variations could be explained as statistical 're-shuffling' of factors during the reproductive process. Variations were never acquired directly through the environment, but were a consequence of intracellular processes. Now, Haldane continued, Bateson was the champion of such a hypothesis, which, if analysed properly, entailed absurd consequences: it implied that all possible factors giving rise to all variations during evolution had to be contained in the first unicellular organism that appeared on earth. '[In the] presence of such hypotheses', concluded Haldane, 'one feels that it is high time to return to sanity'.[55]

One year before his death in 1936, when his famous son J. B. S. Haldane, together with other biologists, was involved in a movement that aimed to synthesize Mendelian genetics, Darwinian evolution and other biological disciplines, Haldane the father left his last reflections on heredity in one of his most 'Romantic' works, *The Philosophy of a Biologist*.[56] Here Haldane supported the possibility of acquired characters as the origin of novelty in evolution; the passage is worth quoting in full:

> In the acquisition of a new character the organism is just as much active as the environment, and were it not so new characters would not be transmitted to descendants. It is only in the light of the distinctively biological conception of a life which embraces environment that evolution can be interpreted. Every new character is an active adaptation of pre-existing life, and its transmission to descendants is a sign that in the adaptation the life itself of the organism is expressing itself ... In the same way the attribution in individual development of all adult characters to the physical and chemical structure of its germinal material assumes that we can describe life as appertaining to physical and chemical structure apart from the environment. This we certainly cannot do. *The chromosomes and chromomeres of a living germ-cell are an expression of its whole life, and its further life in essential connexion with embryonic and adult environment furnishes the only key to an understanding of their real nature.*[57]

What is really striking is that Haldane never mentioned Lamarck; however, his view on evolution was entirely Lamarckian. Organisms constantly adapt themselves to the environment, and in so doing, they improve. In transmitting their 'ameliorations' to descendants, organisms constantly become more complex. In short, evolution, in the eyes of Haldane, was characterized by constant progress, and human society was not excluded from such a scheme.

## The Organism as a Metaphor of Society

Haldane's vision of biology, shaped by organicism (that, in part, he took from Delage) and Lamarckian evolution, was applied to his views on society.[58] As his son J. B. S. recalled, although Haldane had never been a radical, he was very concerned about social conflicts due to the iniquitous distribution of wealth. When, in 1883, Haldane published the previously mentioned article with his brother Robert, he attacked what he considered 'the idea of the state as mere aggregate of isolated individuals'. Indeed, they continued, 'A less abstract category would prove more adequate to the facts in embracing, in the conception of the individual, his determination by the social organism of which he is a member'.[59] In the following years, in echoing Hegel's *Philosophie des Rechts* – where the relation between organism and society was explicitly articulated – Haldane expressed his political convictions and ideas of society during his numerous public lectures to societies and institutions.[60] For example, in 1924, in a period of social unrest and communist threats, Haldane was elected president of the Institution of Mining Engineers. In the presidential address entitled 'Values in Industry', he exposed his ideas and convictions about the aims and scope of the institution in regard to British economy and society. First of all, engineers should always have a special care about safety and health in the working conditions of miners and employees. The mining management should give the highest value to human comradeship, and in so doing, it had to promote education and housing for all miners and their families. Miners must be considered precious fellows and not mere unskilled workers:

> It is not with scientific abstractions called 'labour' or 'capital' that British mining engineers have to work, but with their own fellow countrymen, their own flesh and blood. These fellow-countrymen will give loyal and efficient service, will face any danger, will forgive imagined or real mistakes, and will take the rough with the smooth, the bad times with the good; but what they will not tolerate is being treated as if they were mere tools, to be cast aside without compunction.[61]

Miners must be considered people and not gears in a capitalist machine. A colliery, like society, had to be seen as an organism, with all its different parts unified in a superior, unique whole. Indeed, if a company – an industrial enterprise – were deemed a mere economic machine, where its employees represented selfish units or independent individuals, then 'it has no soul to hurt, though it has body to kick'.[62] However, Haldane added, if we consider society and industry as mere aggregates of selfish individuals pursuing their own interests, we should not be surprised by social unrests and uprisings. The struggle between classes, between employers and employees, between capitalists and the working classes, was not only due to a mere unequal distribution of goods, but was also a 'human rebellion against what are regarded as inhuman relations'.[63] In a colliery, for example, if positive relations could be established, a natural cooperation and comradeship would follow. The real problem – indeed the main cause behind strikes and revolts – lay in a materialist and mechanist conception of economy which ran against the natural predisposition of men to cooperate for a unique goal. The concrete human being, as seen in his working and social context and in his daily life, was not, as depicted by economists, a selfish individual, but a person naturally endowed with sympathies towards his fellows: human communities worked as whole organisms, and when solidarity among parts was disrupted, the social organism (community) died.

Haldane had no sympathies for communism; he was rather a liberal socialist, thinking that societies could progress and thrive as long as the relations among their members were 'human' relations. Haldane's liberal thinking was directly inspired by T. H. Green, an idealist philosopher and political radical who proposed an influential form of ethical socialism.[64] Societies, like living organisms, were purposeful entities where parts and whole were in constant, irreducible relation.[65] From his Romantic conception of life, a conception based on his physiological findings as well as his philosophical convictions, Haldane not only criticized medical sciences and current conceptions of heredity, he also applied organismal notions to British capitalism. As Sturdy remarked: 'Haldane saw the demonstration of biological teleology as a way of vindicating his metaphysical world view, with all the implications he believed that held for social theory'.[66] Haldane thought that idealist and organicist philosophies were strictly related to progressivist social reforms. Refuting materialist and mechanist sciences did not mean rebutting an abstract position; it meant refuting a whole political ideology that made men slaves of a mechanic state. With Haldane, twentieth-century Romantic biology acquired a leftist tint.

## A Romantic Morphologist

D'Arcy Thompson shared most of Haldane's universe. Unlike Haldane, though, Thompson has been portrayed as an idiosyncratic, heretic and unorthodox naturalist unconnected with his contemporary science. Although S. J. Gould recognized that Thompson's critiques against neo-Darwinian evolution were shared by many zoologists during the first decades of the twentieth century, he maintained that Thompson's science of form clashed with mainstream evolutionary biology. As Gould explained:

> If this general and multifaceted critique of Darwin functionalism ... harmonized well with the major trends of thought in the evolutionary biology of his generation, the second prominent implication that he drew from his idiosyncratic theory of form could not have stood in more oppositional relationship to an even deeper, and even more general, assumption of the evolutionary science, both in D'Arcy Thompson's day and in our own.[67]

As I will show in this section, this is partially untrue. Gould's interpretation, like other readings seeing Thompson as isolated and unconnected,[68] overlooked the real network of scientists that supported or simply inspired Thompson's agenda, i.e. developmental biologists, morphologists, comparative anatomists, physicians and biochemists. Thompson's science of form was not an alternative to evolutionary sciences; it was an alternative theory to neo-Darwinian evolution. Together with Haldane and many twentieth-century 'Romantic' naturalists, Thompson's proposal was inscribed in a well-defined research programme in developmental evolution: evolutionary diversification was not due to stochastic gene mutation and selection but to specific changes of growth and form happening during individual development. To him – along with many of his contemporaries – mere historical explanations were suspect: they appeared as controversial and unsupported speculations. Indeed, any explanation about organic evolution had to start with knowledge of the laws ruling individual development. Differentiation of the organic form could be explained not with Darwinian evolution but with individual growth rates emerging from a specific 'system of forces'. This was, as we will see further on, a staple theme among organicist zoologists.

Thompson's science of form was therefore not the product of an idiosyncratic and isolated mind. He had many friends and admirers that pursued, or simply crossed, his agenda. As Thompson's archive in St Andrews demonstrates, he was in contact with many well-known or lesser known figures supporting, advancing or simply responding to his investigations. Behind Thompson's seminal work *On Growth and Form* (1942), there was a vibrant network of scholars sharing his insights and methods, including, among others, A. Dalcq, H. Przibram, E. Hatschek, E. Conklin, J. Huxley, F. Lillie and, of course, Haldane. Dalcq (1893–1973) was a young embryologist working at the Université libre de Bruxelles. Trained by A. Brachet (1869–1930), Dalcq had published an important book with Cambridge University Press: *Form and Causality in Early Development* (1938).[69] Dalcq admired Thompson's works, and in 1939 he sent a letter addressed to the Secretary of the Royal Society, asking for a portrait of the Scottish professor: 'When I was a young assistant', Dalcq explained, 'I once saw Professor D'Arcy Thompson and attended here, in Brussels, his memorable lecture. This was for me an impression of which I now realize it has influenced my whole research life'. As he further remarked, Thompson was the 'real promoter of the mode of interpretation which is becoming prevalent in the problem of early morphogenesis'.[70] In several enthusiastic letters Dalcq addressed to Thompson, he not only expressed his admiration for the Scottish scholar but also invited him to Brussels to deliver seminars and lectures. Indeed, as Dalcq mentioned, at the ULB's Faculty of Medicine there were scholars sharing most of Thompson's interests and working on similar lines of research.[71] Dalcq himself was well in line with Thompson's holistic understanding of biological sciences. After all, he aimed to provide a *causal* physico-chemical explanation of organic form and development.

H. Przibram (1874–1944) had graduated in zoology at the University of Vienna under the supervision of B. Hatschek (1854–1941), a morphologist and embryologist who had been a student of E. Haeckel and R. Leuckart.[72] Przibram's agenda included mathematical studies on organic morphogenesis; in particular, he aimed to understand the physical forces shaping the organism's development through a quantitative approach encompassing physical, chemical and physiological knowledge. No wonder that Przibram's project for a mathematical morphology had many similarities with Thompson's approach. Przibram deeply admired Thompson and his work. In a postcard he even called Thompson his 'Dear and venerated Master!'[73] Thompson and Przibram learned from each other about mathematical morphology and other issues in the life sciences.[74] As Cohen recounts, 'Research at the Vivarium provided D'Arcy Thompson himself with inspiration for his mathematical morphology'.[75] Przibram, with Thompson, believed that physics and mathematics could be excellent allies for life sciences: they would provide a deeper understanding of the organic form, its development and its evolution. Although Thompson and Przibram did not always agree about the applications and approaches of their mathematical morphology, they deeply respected each other's work.[76] When, in 1928, Przibram was invited to give a lecture at the Royal Society of Edinburgh, he wrote to Francis Crew (1886–1973) and made clear that he wanted to meet D'Arcy personally.[77] In 1944 the Viennese zoologist was killed in the Nazi concentration camp of Theresienstadt in the Czech Republic, and Thompson wrote a moving obituary in his memory.[78] In the second edition of *On Growth and Form*, Thompson made large use of Przibram's findings.

E. Hatschek (1868–1944) was a Hungarian émigré living in London since 1888.[79] He had studied chemical engineering at Zurich and Vienna. In England he became a notorious scholar in colloidal chemistry. As an independent researcher, he published several original and introductory works in the field; in particular, his *Introduction to the Physics and Chemistry of Colloids* (1913) went through five editions.[80] The correspondence between Hatschek and Thompson is important both quantitatively and qualitatively. Hatschek was interested in issues related to the relation between surface tension, energy, viscosity, vibrations, strains and shapes in gelatinous substances. With Thompson, he believed that there were interesting relationships between the physical behaviour of inorganic substances and the organic form. One year before publishing the first edition of *On Growth and Form*, Thompson explained to Hatschek: 'I feel all but certain that the whole symmetry of the coelenterate ... of the part of the coral and of the jelly-fish are simply referable to an unseen vibration going on within the colloid substance'.[81] Hatschek encouraged Thompson to perform experiments. In particular, he sent specimens of gels with instructions attached for handling the materials and getting organic forms: 'I send you separately the first specimens of imitations medusoids or zoophytes ... which I made after very little trying. They consist of 20% gelatin, stained with indigo-carmine, which is dropped into a coagulating solution ... if the specimens should arrive unrecognizable, you will have no difficulty in repeating the experiments'.[82] Hatschek took Thompson's agenda very seriously. He tried to reproduce different kinds of organic forms using all his chemical expertise.[83] Thompson was delighted with Hatschek's successful outcomes, though he lamented that many zoologists did not get the importance of these results:

> It seems to me that none of these zoologists grasp the points. They look upon the things as mere 'curiosities'. They fail to grasp the point, viz. that we have shown, beyond a doubt, that the form of Medusa is a figure of equilibrium under certain peculiar conditions, and that therefore it is not to be explained as something attained, by way of 'adaptation', and by means of slow gradations of evolutionary development.[84]

The several letters that Hatschek sent to Thompson were long and detailed receipts of how to reproduce animal forms with simple chemical substances. Hatschek's meticulous experiments sustained one of Thompson's dearest ideas: that animal form was not the result of natural selection, but was the outcome of determined physical forces.[85] Many of the observations gathered by Hatcheck were later used by Thompson throughout the second edition of *On Growth and Form.*

Thompson was also in contact with a well-known American cytologist and organismal zoologist. E. G. Conklin (1863–1962). Conklin deeply admired Thompson's work and erudition. Conklin had met Thompson in Edinburgh in 1936, and he never forgot that event: 'I remember with the greatest pleasure our first meeting in Edinburgh in 1936 and your later visit to us in this country. I wish we might get together again very soon'.[86] They used to send each other their works. When Conklin received a copy of the new edition of *On Growth and Form* in 1942, he was deeply moved. His enthusiasm betrays, I think, the importance and fame of Thompson among American scholars. Thompson conserved Conklin's reply in his own copy of *On Growth and Form*:

> I cannot adequately express my thanks for this treasured gift with its gracious and affectionate inscription to me ... I have dipped into it here and there and I am ready to say of it what Emerson said of Plato: 'in this are all things, whether written or thought'. It is an amazing work of science and philosophy, of literature and art, of learning and wisdom, and I think it could not have been written by any other man of this age than yourself. It is Gargantuan in scope and execution and a model of genuine science and scholarship, but it also has its appeal to the homely experiences of plain people ... I am pleased that you leave room in your mathematic and mechanics, in your science and philosophy, for ends as well as means, final causes as well as efficient ones. Perhaps ends are not causes and 'final causes' a misnomer, but I am more than ever convinced of a fundamental teleology in nature running alongside of, but not interfering with, causality. The relation of mechanism to finalism seems to me to be unlike that of structure and function, – two complementary aspects of life.[87]

F. R. Lillie (1870–1947) was the director of the zoology department at the University of Chicago and president of the Woods Hole Marine Biological Laboratory.[88] Back in the 1930s Lillie was a very famous developmental biologist, and Thompson particularly admired his works on the physiology of feather development. In 1942 Thompson sent Lillie an early copy of the new edition of *On Growth and Form* and wrote, 'It is a loss to me that we have never met, and a matter of great regret that we are little likely to do so ... Pray accept the new edition of my book on *Growth and Form*, of which I am sending you an early copy today, under separate cover. I am sending it you for what it may be worth, to bridge the gulf between us'.[89] Lillie replied with enthusiasm, betraying his admiration for Thompson's undertakings: 'Your kind letter tempers a little the regret I have long felt at not having met you in the flesh. Your beautiful translation of Aristotle's *Historia Animalium* has long been one of my joys; and the speaking acquaintance that I have had with your *Growth and Form* renders your gift of the new edition, now on the way, an appreciated honour'.[90] The gulf was indeed bridged, and in the following months of 1942 they started a correspondence discussing feather development as well as other more mundane topics.[91]

However, one of the most important figures in Thompson's large network was surely J. Huxley (1887–1975). The friendly relationship between Thompson and Huxley is particularly important as they learned from each other and explicitly admitted that their science would not be the same without their reciprocal contributions. Huxley had been trained in zoology at the University of Oxford, and after a brief period of research in Germany, he took a position as demonstrator at Oxford's department of zoology and comparative anatomy.[92] Huxley started correspondence with Thompson in 1925 when he was appointed Professor of Zoology at King's College London. The dates are not casual. From the beginning of the 1920s, Huxley was interested in issues related to relative growth in fiddle crabs and other organisms.[93] In particular, Huxley's attention was focused on the morphological correlation of changes happening between the parts and the whole during an organism's growth. As is well known, in a paper published in 1924, Huxley formulated the equation describing the law of heterogenic growth: $y = bx^{\alpha}$. Yet in 1936, together with G. Teissier, he coined the term 'allometry'.[94] All that concerned relative growth was of high interest to Thompson. Indeed, in 1925 he sent a letter to Huxley which stated, 'I have been looking at your last letter ... in which you gave me G.H.H.'s formula, for the rates of growth of x and y, at rates p and q, and the final value of $y = bx^{\alpha}$ ... I can see that this work of yours is going to improve vastly my Growth-chapter in the next edition'.[95] However, Thompson firmly believed that Huxley was building on his own original work, and improving or extending his own novel results. Indeed, concerning Huxley's law of 'constant differential growth', Thompson felt that in the first edition of *On Growth and Form* he had already expressed it, though implicitly:

> I don't care a hang (I have perhaps cared all my life too little) about priority; but, since writing you last night, I have been wandering how far I expressed your idea of a 'constant differential growth-ratio', which was certainly present to my own mind all along ... I don't want to deny that you now put it much better; but still ... I think I had precisely your own main points clearly in mind. I don't want to question, or detract from, your own originality, your own independent discovery, in the very least; but I do want to feel satisfied that I was not so blind to an elementary and fundamental point as you perhaps think I was.[96]

Huxley never questioned the pioneering work of Thompson. In 1932, when Huxley published *Problems of Relative Growth*, he let Thompson know that the book was dedicated to him: 'Here at least is an advance copy my growth book. You remember I asked you if I might dedicate it to you, and you said you would be glad for me to do so. I hope you will feel that it is a worthy offspring of *Growth and Form*'.[97] However, Huxley not only began the book with a long quotation from Thompson, but also introduced the book by stressing his debt to the Scottish master: 'I owe a great deal to previous work in the field; first and foremost to D'Arcy Thompson's *Growth and Form*'.[98] Huxley also helped Thompson to improve the new edition of *On Growth and Form*, providing comments and advice.[99] Indeed, one year before Thompson published the second edition in 1942, he sent a draft copy of it to Huxley. Although after World War II Huxley's interests were radically changed, he was extremely pleased: 'How very kind of you to send a copy of your book! I think a good deal of it ... You are indeed unique with your combination of different branches of learning'.[100]

Once we realize that Thompson and his scientific world were well rooted in the organicist tradition, connected to an international discussion on the nature of life phenomena, well established in a traditional setting of studies in developmental biology and evolution, and well founded in disparate international institutions and laboratories, we have to revise his bio-philosophy, as well – in particular, his views about reductionism, mechanism and teleology.

## Against Reductionism

Thompson began his *magnum opus*, *On Growth and Form*, with Kant: 'Of the chemistry of his day and generation, Kant declared that it was a science, but not a science – *eine Wissenschaft, aber nicht Wissenschaft* – for that the criterion of true science lay in its relations to mathematics'.[101] Of course, the Kant approached by Thompson in these first lines was not the Kant of *Critique of Judgment* but the very different Kant of the *Critique of Pure Reason*, namely, a Kant enamoured of Newton's physics, Euclid's geometry and Laplace's mathematics. Not surprisingly, Gould characterized Thompson as 'physicalist',[102] while Barry Ninham and Pierandrea Lo Nostro labelled the Scottish professor as 'reductionist'.[103] Some see Thompson as a precursor of the methods employed by biochemists, and others as a predecessor of molecular biology.[104] Passages of his works can plainly justify these interpretations. His crusade against different forms of vitalism and teleology and his defence of mechanist approaches in life sciences made him, apparently, a good fellow for the 'physicalist' campaign. However, to see Thompson's biology as essentially physicalist would be misleading. In fact, even though a modern reader might see Thompson's emphasis on the importance of mathematics in the study of living things – and the weight he gave to physics and mechanism to explain organic forms, as well as the importance he attributed to geometrical symmetries in morphology – as a merely reductionist or physicalist approach, it was not so intended by him and by many of his contemporaries. In the first decades of the twentieth century, the use and application of mathematics to solve biological problems was considered a very promising strategy for many 'holist' theorists; the existence of some simple geometrical forms and quantifiable physico-chemical processes was not in contradiction with organicist biology, and the mathematization of morphology was consistent with teleological approaches or stances.[105]

There is one fundamental distinction we should draw before I introduce Thompson's philosophy of the organism and biology: it is one thing to argue that living bodies are totally explainable and completely reducible to physico-chemical properties, and another to say that organisms never contravene physical laws or disprove chemical knowledge. Thompson's position leaned towards the latter and was widely shared among organicists; he believed that organisms did not break any physical law, and they did not require any vital principle or transcendent energy in order to work, apart from physics and chemistry; however, he felt that organisms exhibit some peculiarities that, as George Kimball nicely explained, made living entities special: 'Thompson tries as often as possible to show the identity, or at least the analogy, between ordinary particles and organic creatures ... and yet he considers that there is a unity of the living body which the non-living does not have. The unity is sometimes conceived as subservience of organic part to the whole, sometimes as an absence of dividing line between head and body – i.e. a physical continuity of material'.[106] Although the difference between organic and inorganic matter was not seen by Thompson as fundamental, nonetheless living bodies manifested different mechanical properties due to specific chemical changes.[107] Thompson was an organicist; he had never been a reductionist. As Peter Medawar clarifies, Thompson tried to integrate different approaches and not merely to reduce life to physico-chemical laws:

> We are mistaking the direction of the flow of thought when we speak of 'analysing' or 'reducing' a biological phenomenon to physics and chemistry. What we endeavour to do is very opposite: to assemble, integrate, or piece together our conception of the phenomenon from our particular knowledge of its constituent parts. It was D'Arcy's belief, as it is also the belief of almost every reputable modern biologist, that this act of integration is in fact possible.[108]

## Teleomechanism

As is well known, Thompson criticized the overuse of teleological thinking in biology. However, he was too Aristotelian to deny purpose in nature.[109] What he mainly denied was Darwinian teleology or, to put it differently, the adaptationist thinking according to which any organic conformation was there because it was useful for the survival of the organism during its phylogenetic past. Along with many biologists belonging to his network – i.e. Haldane, Lillie, Conklin, Przibram, Woodger and Russell (see Chapter 4) – Thompson felt that this adaptationist thinking introduced a speculative approach into biology – an approach very hard to disprove because it supposed what it pretended to explain. Organisms were nicely adapted because they had been selected for, and, as Medawar complains, explaining Thompson's position, 'the formula would accommodate all comers'.[110] In addition, the hypothesis that adaptation was explainable through the mechanism of natural selection could not be accepted because natural selection was considered, at most, a negative factor in evolution.[111] It destroyed but never created new organic forms and, therefore, new adaptations.[112] In brief, not only could natural selection be considered, at most, a very secondary mechanism of evolutionary diversification, but it also introduced a bad form of teleology.[113]

There is probably no better chapter than that which Thompson dedicated to form and mechanical efficiency to illustrate the tension between bad and good uses of teleology in biological thinking: whereas explanation based on adaptation through natural selection mirrored an empty theological reasoning (not very far from Paley's style), the perfect adaptations that organisms manifested could be explained with the use of analogies taken from architecture or human artefacts. In these, the efficient and final causes (to use Aristotelian terms) are evident; efficient causes subsist for the sake of final causes, though the former are comprehensible only assuming the latter. In other words, the good use of teleology must be always associated with efficient causes – one explains the other, and vice versa:

> The naval architect learns a great part of his lesson from the stream-lining of a fish; the yachtsman learns that his sails are nothing more than a great bird's wing, causing the slender hull to *fly* along; and the mathematical study of the stream-lines of a bird, and of the principles underlying the areas and curvatures of its wings and tail, has helped to lay the very foundations of the modern science of aeronautics.[114]

Human artefacts are intrinsically purposive, as both organisms and their parts are made to work under precise circumstances and operate according to determined physical constraints. Just as bridges built by engineers are designed to resist certain specific loads and stresses, the architecture of vertebrate skeleton is made accordingly. The skeleton of a bison, horse or ox works like a suspension bridge: all are made according to a well-planned distribution of lines of force and tensions. Now, for Thompson, the efficient causes lying behind the skeleton's adaptations included the phenomena of growth: bones grew under specific strains, stresses and tensions, with all factors guiding and shaping the skeleton's architecture as a whole:

> Each animal is fitted with a backbone adapted to his own individual needs, or (in other words) corresponding to the mean resultant of the many stresses to which as a mechanical system it is exposed, [therefore] ... skeletal form, as brought about by growth, is to a very large extent determined by mechanical considerations, and tends to manifest itself as a diagram, or reflected image, of mechanical stress.[115]

However, mechanical stresses – as the efficient causes explaining bone morphologies and skeleton structures – were comprehensible only in the light of assumed functions. Unlike bridges, the skeleton form was the product of a progressive growth shaped from functional stimuli. Strains, stresses and tensions derived from the specific conditions of life during development. The particular structure of an organism was intrinsically correlated to its life activities; in other words, walking, running, jumping or flying – activities performed ever since the first stages of the organism's life – shaped the entire morphology. The organism was a continuous 'work-in-progress', a system that was simultaneously active and passive to the environmental stimuli. This meant that efficient and final causes had to be considered as two faces of the same coin. Thompson followed Goethe, who clearly argued in his *Die Metamorphose der Tiere* that in organisms, 'none of their functioning members are ever opposed to the others but labor in common toward life. Thus does its form determine an animal's manner of living and manner of living in turn react upon every form in the most powerful way. Thus the organized being is stable'.[116] The same discourse was readily applicable to other cases: the forms of *Spicule* or *Foramifera*, leaf arrangements or cellular aggregates. Nature was filled with instances showing the close interdependence of efficient and final causes. So it is no wonder that Thompson opened his *On Growth and Form* with the following words: 'like warp and woof, mechanism and teleology are interwoven together, and we must not cleave to the one nor despise the other; for their union is rooted in the very nature of totality'.[117]

Although Thompson was not a reductionist and never accepted pure structuralist biology, he believed that physical and chemical sciences offered the best tools for studying and understanding organic form. Cells, tissues and organs, followed by flowers, leaves and trees, all obeyed physical laws.[118] First of all, as he himself admitted, even though, as a hypothesis, he considered the organism as a certain kind material and mechanical entity,[119] he was well aware that physico-chemical properties were often not enough to explain life phenomena: 'One does not come by studying living things for a lifetime to suppose that physics and chemistry can account for them all'.[120] In total agreement with Lotze's conception of 'teleomechanism', Thompson deemed mechanical and physical representations to be useful simplifications, heuristic tools and promising conceptual vehicles in understanding the simplest manifestations of life. Physical mechanisms were instruments through which organic purposes realized and manifested themselves. Thompson never betrayed Kant's idea, according to which the student of living things should stretch, as far as he could, mechanical explanations to life, although the natural philosopher should be aware that those explanations cannot go very far where life is concerned. No wonder that Thompson's biology was essentially focused on the lower organisms and the simplest of life's manifestations.

## Beyond Genetics: The Organism as a Whole

> if the cells acts, after this fashion, as a whole, each part interacting of necessity with the rest, the same is certainly true of the entire multicellular organism ... as Goethe said long ago, 'Das lebendige ist zwar in Elemente zerlegt, aber man kann es aus diesen nicht wieder zusammenstellen und beleben'.[121]
>
> D'Arcy W. Thompson

The particular form of organicism that Thompson defended surrounded the central theme of *On Growth and Form*: how different kinds of forces shaped organic forms. The book follows a Bergsonian theme. It consists in a long list of cases showing how forces 'compenetrate' organic matter in directing (or constraining) growth according to specific trends, therefore producing diverse configurations. From the smallest organisms (for instance, minute bacteria, human blood corpuscles and protozoa) to middle-sized creatures (for example, minute insects, jellyfish or tiny fish) and subsequently to the largest life forms (for example, mammals or birds), all were influenced by diverse forces: surface tension in the first, surface tension and gravity in the second, and gravity in the last case. Of course, growth was not only constrained by direct physical forces: it was a dynamic process entailing differential rates among various organic parts. The rates depended on osmotic, catalytic and physiological secretions that, interacting with extant temperatures, retarded or accelerated the growth of organisms. For Thompson, as long as growth rates changed, the overall configuration of the organism changed accordingly.

The long and complex argument, explicated through more than one thousand pages, was not the outcome of an idiosyncratic mind: it was the result of a broad range of national and international allies, both ancients and contemporaries. When Thompson published the second extended edition of *On Growth and Form*, he upgraded the new version with a reference to Lillie's works on the physiology of feather development. Yet Lillie was not the only resource on whom Thompson relied and used in his book. Another important American figure represented for him a privileged source of inspiration: C. M. Child (1869–1954). As a colleague of Lillie in Chicago, Child was a well-known physiologist interested in animal regeneration, and Thompson's interest in Child is not surprising because the phenomena related to animal regeneration were seen as relevant instances of organic growth in general.[122] In particular, he was fascinated by Child's experiments on planarian regeneration, which demonstrated the existence of gradients following the body axis of the organism: gradients operating between two poles, one dominant and the other subordinate.[123] With Child, Thompson felt that the configuration of an organism was a function of its differential growth and, as such, it characterized a space-time event and not a mere spatial entity. With Child, Thompson assumed that differential growth rates depended on several factors, particularly age and external temperature. Finally, and most importantly, with Child, Thompson concluded that differential growth rates are behind all morphological characters, peculiar or shared, of all organisms, and therefore specified individual or species qualities and conformations.[124] In sum, like Child, Thompson was saying that morphological variations were due to growth variations rather than genes or any other posited nuclear particle.

After discussing growth rates and regeneration, Thompson moved to cell theory. This is a chapter often overlooked by previous interpreters. However, it represents – together with the chapters on form and mechanical efficiency and on the theory of transformations – the best evidence proving that Thompson was irremediably part of the organicist tradition. First of all, he underscored de Bary's and Whitman's critique of cell theory:[125] cells were not the individual building blocks of the organism because the organism itself coordinated and drove their distribution. Famous naturalists, botanists, zoologists and embryologists of different periods were enlisted by Thompson to support this stance: as well as de Bary and Whitman, he also mentioned Goethe, Sachs, Rauber, Wilson, Sedgwick and others. All together, they strongly reminded that in cellular differentiation 'we deal not with material continuity, not with little bridges of connecting protoplasm, but with a continuity of forces, a comprehensive field of forces, which runs through the entire organism and is by no means restricted in its passage to a protoplasmic continuum'.[126]

Thus the entire organism was conceived as more than its composing parts; however, what about individual cells? The position Thompson defended was, once again, the standard stance supported by most of the organicist biologists. In particular, drawing on John Goodsir and Bergson,[127] Thompson saw the cell as a *sphere of action*, where an unstable equilibrium of forces dwelled.[128] Surface tension, of course, was present, but chemical and electrical forces were also involved. Yet the cell is inseparably connected with its environment.[129] A dynamical conception of the cell, Thompson maintained, must entail that the whole system work as an integrated individuality: there were neither active or passive regions, nor relevant or irrelevant parts, 'for the manifestations of force can only be due to the *interaction* of the various parts ... certain properties may be manifested, certain functions may be carried on, by the protoplasm apart from the nucleus; but the interaction of the two is necessary, that other and more important properties or functions may be manifested'.[130] In fact, for Thompson and many other contemporaries upholding the importance of the cell surface and cytoplasm for reproduction and inheritance, the cortical regions of the cell played a fundamental role in all cellular activities.[131]

Thompson's organicist conception of the cell is particularly important when we consider his critique of Weismann's germ plasm and Mendelian genetics; even though he was not directly opposed to the latter, it certainly did not fit his general framework.[132] The reasons for that derived not only from Thompson's views on the cell, but also from his understanding of cytology. Firstly, Thompson believed that investigations about cell functions were more important than knowledge about cell structure. In fact, studying cell structure did not necessarily advance one's knowledge about cellular activities: 'The *things* which we see in the cell are less important than the *actions* which we recognise in the cell'.[133] To Thompson, identifying specific structures – for instance, the germinal spot, chromatin or chromosomes – did not explain, *de facto*, cellular functions. All hypotheses that correlated special properties to the matter contained in the cells – hypotheses such as Darwin's pangenesis or Weismann's germ plasm – confused the action of energy with supposed physical particles: 'If we speak, as Weismann and others speak, of an "hereditary substance" ... we can only justify our mode of speech by the assumption that a particular portion of matter is the essential vehicle of a particular charge or distribution of energy, in which is involved the capability of producing motion, or of doing work'.[134] All particulate theories of inheritance committed the sin of attributing to matter what was a manifestation of cellular forces; inheritance was not only a property of the whole cell, it was also a manifestation of energy that, for its own nature, could not be identified or circumscribed to specific cell areas.[135] Although all this ran against Darwin's and Weismann's hypotheses of heredity and development, it could not be easily fitted with the new paradigm developed by Morgan and his students at the Columbia Lab because, as we will see clearly in the next section, for Thompson, heredity was not based on the transmission of 'factors' or 'genes' but on the transmission of systems or equilibriums of forces.

Even though Thompson rarely mentioned heredity and never developed a personal and coherent view on the field, his ideas on how organisms transmit and maintain their form, generation over generation, were clearly expressed when he dealt with evolution. Indeed, to him evolution was not only Lamarckian in principle (although an adulterated form of Lamarckism), it was also based on a holistic perspective of evolutionary transformations and novelties, transformations entailing whole and dynamic systems of energy which could never be reduced to some physical properties of the cell nucleus alone.[136] As Thompson claimed in 1923, 'The chromosome people are having a good innings; but their theories are top-heavy, and will tumble down of their own weight. It is of little use, meanwhile, to argue with them'.[137]

## Holistic Evolution

> The biologist, as well as the philosopher, learns to recognise that the whole is not merely the sum of its parts. It is this, and much more than this. For it is not a bundle of parts but an organization of parts, of parts in their mutual arrangement, fitting one with another in what Aristotle calls 'a single and indivisible principle of unity'; and this is not merely metaphysical conception, but is in biology the fundamental truth which lies at the basis of Geoffroy's or Goethe's law of 'compensation' or 'balancement of growth'.[138]
>
> D'Arcy W. Thompson

In 1929, as I remarked earlier, Thompson and his friend Haldane attended the British Association conference in South Africa, where they discussed holism and evolution with Smuts. They shared many ideas about evolutionary biology: with Haldane and Smuts, Thompson believed that evolution had nothing to do with transmission and selection of discrete characters because there was nothing purely discrete in the organic world. It is revealing that Thompson directly quoted Johannes Müller in rejecting what he called the doctrine of single characters, i.e. the idea that the adult organism represents, in truth, a set of discrete hereditary factors. 'Die Ursache der Art der Exitenz bei jedem Theile eines Lebenden Korpers ist im Ganzen enthalten' was indeed a statement that Müller, in his celebrated *Handbuch der Physiologie des Menschen* (Textbook of Physiology), attributed directly to Kant.[139] It conveyed the idea that not only was the organic whole explicatively more important than its parts, but also that organic parts could only be separated through a process of abstraction – biologists' distinctions – so that organs, tissues, cells, bones, etc. were mere theoretical tools that could be helpful only if the scholar was fully aware that they were, in truth, abstractions. Organisms, from an ontological viewpoint, were irreducible and composite wholes, and evolutionary mechanisms had to operate on functional wholes and not on small, discrete characters.[140]

Now, for Thompson, one of the most evident instances of organic integration was the animal skeleton and its efficiency; it showed that the old morphological and anatomical observations – observations based on the structural analysis of parts as abstracted from the whole – were necessarily misleading.[141] In fact, mirroring Kant's saying, he argued that the same principles were at work both in the architecture of the whole skeleton and in the constitution of a single bone. Yet the skeleton and bones were in turn 'moulded' with muscles and other tissues so that a complete understanding of the skeletal structure and function was possible only if the whole were taken into account. Even though he believed that organisms could be seen as functional wholes, he preferred to focus on the idea that living beings were a product of well-defined physical causes acting on the whole organic system of forces. This is the central idea behind his hypothesis of transformation. In the same way as the skeleton was a result of pressures, stresses and forces acting on growing heterogeneous matter, so too the forms of organism were the product of forces acting on the whole plant or animal structure. If forces explained form, then heredity and evolution (as conceived by Thompson's contemporaries) lost its privileged explicative power: 'To look on the hereditary or evolutionary factor as *the guiding principle* in morphology is to give to that science a one-sided and fallacious simplicity'.[142]

Indeed, to Thompson, although affinities among organisms (and morphological similarities among their parts) could indicate phylogenetic relations, they could also be (and often were) the product of similar forces acting on the system as a whole. After all, organisms can be similar to each other or maintain their invariable morphology for many generations because they live in similar environments and, therefore, are subjected to similar stresses, pressures and forces. If form were strictly related to growth rates (which Thompson dubbed morphological heredity), and if growth rates were related to environmental stresses and forces, organisms' morphology may change if environmental forces change. In other words, the systems of forces constraining or influencing growth rates can be disrupted by external forces or new habits, which in turn may lead to a new equilibrium or system of forces. As a consequence, in evolution organisms were transformed neither through acquired *characters* nor through stochastic *gene* mutations; they changed through new 'acquired' systems of forces during organic development:

> The deep-seated rhythms of growth which, as I venture to think, are the chief basis of morphological heredity, bring about similarities of form which endure in the absence of conflicting forces; but a new system of forces, introduced by altered environment and habits, impinging on those particular parts of the fabric which lie within this particular field of force, will assuredly not be long of manifesting itself in notable and inevitable modifications of form.[143]

The system of forces Thompson had assumed was represented in the last celebrated chapter of his book: they were pictured as Cartesian deforming grids outlining the *bauplans* of organisms. Similar and related forms were compared and described through these geometrical grids. Thompson thought that similar forms varied in correspondence with various deformations of these coordinates, representing lines of force acting on the organism's whole body and on its parts.

Unlike the old morphological tradition which tended to analyse and compare single morphological parts, Thompson's diagrams did not illustrate variations or comparisons of single characters, but an overall reshaping of the entire *bauplan*. After all, he remained loyal to Cuvier, who had argued that 'correlation' among characters was a distinctive feature of the organism. Specific morphological characters were once again considered by Thompson as a mere abstraction, only conceived in the mind of the investigator, because, as he claimed:

> The living body is one integral and indivisible whole, in which we cannot find, when we come to look for it, any strict dividing line even between the head and the body, the muscle and the tendon, the sinew and the bone. Characters which we have differentiated insist on integrating themselves again ... The coordinate diagram throws into relief the integral solidarity of the organism, and enables us to see how simple a certain kind of correlation is which had been apt to seem a subtle and complex thing.[144]

Therefore coordinates illustrated organic 'holist' transformations as represented in closely related *bauplans*. Indeed, Thompson's grids were not appropriate to deal with transformations of unrelated or barely related *bauplans* because, just as in geometry, it is impossible to conceive a transformation of very different figures. Thompson used the example of the incommensurability of helicoid and ellipsoid figures: 'We cannot transform an invertebrate into a vertebrate, nor a coelenterate into a worm'.[145] Incommensurability among forms, therefore among certain grids, reflected incommensurability among 'types' in nature, as Cuvier had taught and used as an argument against 'transformationists' in his own time. However, unlike Cuvier, Thompson believed in evolution, and unlike Darwin, he maintained that organic transformation entailed large leaps: 'Our geometrical analogies weigh heavily against Darwin's conception of endless small continuous variations; they help to show that discontinuous variations are a natural thing, that "mutations" – or sudden changes, greater or less – are bound to have taken place, and new "types to have arisen, now and then"'.[146]

With Lamarck and Haldane, Thompson felt that organic forms change over time and evolve as a function of the environment and new acquired habits; however, with Aristotle, Kant, Goethe and Cuvier (and some of his contemporaries), he held that organisms are correlated and integrated individualities. He synthesized all these positions and conceived organic evolution as a process entailing 'holist' transformations among related 'types' – transformations due to inherited and plastic systems of forces.[147] Just as Child had argued that hereditary transmission entailed systems of reaction and not genes, so Thompson believed that evolution required the transmission of systems of forces and not factors. Once again, Thompson was not so isolated. His ideas of evolution were not so heretical among organicist zoologists during the first decades of the twentieth century.

## Concluding Remarks

We have seen how highly relevant the intellectual connections were between Haldane, German science and post-Kantian bio-philosophy. His training in Germany had to reinforce his fascination for such a tradition; his intellectual context – enriched from the discussions with his brother Robert and his lifelong friend D'Arcy Thompson – facilitated his acquaintance with classics in philosophy and science. His direct experiences in the field (and in his own body) strengthened his faith in the Kantian idea that organisms were entities in which 'every part is reciprocally purpose and means'.[148] His important inventions, hypotheses and contributions in physiology were essentially shaped, as he explicitly claimed, by his professed organicism.

Thompson's connections with the Romantic tradition were, though less direct, highly explicit. He grew up and worked in a very similar intellectual context; he admired German biology; and, as a polymath, he had a deep knowledge of its history and its contemporary progress. As a scholar of Aristotle, he never abandoned teleology, even though he attempted to explore possible alternatives. In addition, Thompson's fascination with numbers and geometrical ideals and forms; his quest for simple laws, symmetries and patterns regulating the inorganic and organic worlds (with the help of analogies such as soap bubbles, snow crystals or clouds); his whole conception of Nature as a creative, mysterious, active and feminine entity able to shape and produce beautiful and admirable things; his wonder and mystical respect for all he observed and investigated; and finally, his obsession with organic form: all point the reader – even more than to Hellenic philosophy and Hellenistic mathematics and science – to a personal and original Romantic conception of life.

Yet Thompson was far from being a lonely academic; his work was essentially shaped by the intellectual environment in which he was immersed. His *On Growth and Form* was rooted in the academic network illustrated above. In other words, behind the scenes of that ambitious book, and behind Thompson's discussions about mathematical morphology, growth rates, physical morphology, regeneration and holism, there were Przibram and Hatchek, then Russell, Conklin, Lillie, Huxley, Haldane and Smuts, together with many other lesser known figures, including comparative anatomists, embryologists and morphologists who were against the idea according to which, to use Medawar's expression, 'zoological learning consisted of so many glosses on the evolutionary text'.[149] Haldane, and the scholars constituting Thompson's own network, tended to interpret organic forms as manifestations of direct causal factors and not as a sign of evolutionary relations.[150] Thus Thompson was not an idiosyncratic evolutionist: he was a mainstream developmental biologist interpreting evolution as many embryologists did. *On Growth and Form* was therefore a book widely read and praised precisely because it was understood and interpreted within a developmentalist context – a context where the organism's forms were conceived as the outcome of various kinds of forces acting in diverse ways and magnitudes.

However, Haldane and Thompson had different ideas; they did not share a unique and monolithic set of beliefs. Haldane was a physiologist keen to formulate grandiose philosophical schemes affecting his scientific practices and results, while Thompson was a morphologist trying to apply new mathematical and physical tools to problems that, since Aristotle, had never been solved. What they really shared, though, was Kant's idea that organisms were organized and material wholes, in problematic interaction with their environment, and acting in purposive and creative ways. From diverse perspectives, Haldane and Thompson represented two extraordinary examples of Romantic biology as practiced in the United Kingdom during the first decades of the twentieth century. However, followers of Haldane's and Thompson's physiology and biology did not proliferate; they did not lay the foundations for a school; they did not indoctrinate students. Nevertheless, they were the vanguard of a small crowd that, in the United Kingdom, expanded and updated their insights; E. S. Russell and J. H. Woodger were part of this crowd.

# 4 THE NEW GENERATION: A FAILED ORGANISMAL REVOLUTION

## Haldane's Blessing

In 1931 Haldane published a book representing his Donnellan lectures delivered at the University of Dublin in the previous year.[1] The book addressed some philosophical issues related to practices and theories in biology; in fact, not coincidentally, it was titled *The Philosophical Basis of Biology*. Haldane, now seventy-one years old, not only reported and synthesized all the convictions about philosophy, biology, physiology and psychology he had developed and promoted throughout his intense life; he also felt the need to conclude the book with a brief supplement in which he discussed three recent and promising books: Hogben's *Nature of Living Matter*, Woodger's *Biological Principles* and Russell's *Interpretation of Development and Heredity*, all published between 1929 and 1930. Haldane was rather critical of Hogben's mechanistic interpretation of life phenomena: Haldane complained that in Hogben's reductionist approach (an approach extended to animal behaviour), he overlooked the organic coordination and integratedness exhibited by any organism. Furthermore, his mechanist interpretation of hereditary phenomena – an interpretation drawing heavily on the latest advances promoted by the Mendelian school – supported, rather than undermined, mystical and vitalist hypotheses explaining inheritance. As Haldane thundered:

> I can hardly imagine anything more calculated to make men vitalists of the old school than a contemplation of all the orderly facts relating to the behaviour of chromosomes in cell-division and fertilisation, with the related phenomena of hereditary transmission, together with the fact that we cannot form even the foggiest mechanistic conception of how these phenomena are brought about. To regard them as throwing light on any 'mechanism' of heredity seems to me to be only ludicrous.[2]

Conversely, and not surprisingly, Haldane was quite enthusiastic about Woodger and Russell, who represented a younger generation of scholars that was developing, and even improving, his own insights. Woodger, Haldane declared, 'carries me with him in nearly all his criticism, and his references to my own writings are very friendly, I can only express the hope that his book will be widely read ... His criticisms of the use of both mechanistic and vitalistic ideas are even more thorough, and considerably more detailed, than my own; and he arrives at a conclusion in which I am in entire agreement with him'.[3] Haldane also praised Russell's book, which he considered 'a very scholarly and extremely interesting critical discussion of the theories which have, at various periods in the history of biology up to the present, been held on the subject of heredity and embryology. It is a book which can be very confidently recommended to the careful consideration of all those interested, not only in the subject, but in its general philosophical implications'.[4] Furthermore, Haldane praised the critical discussions that Russell had undertaken about contemporary theories of heredity; Russell's harsh critiques of Weismann's germ plasm and Mendelian genetics were as beautiful music to Haldane's ears.

In short, Haldane, the elder statesman, not only fully supported new and more accurate versions of the theoretical biology he had taught throughout his long career, he also backed and encouraged new generations of scholars in improving what I have dubbed Romantic biology. In this chapter we will see that Russell and Woodger, albeit from different perspectives and interests, took seriously some of the organicist tenets discussed in Chapters 2 and 3: denial of vitalism and mechanism; the acceptance of living organization as a postulate; the use of teleological explanations; the complex dialectic between function and structure; the essential organic relation of parts to the whole; and the fundamental interrelatedness of organism and environment. However, whereas Russell sought only to formulate an original and coherent framework – what he called 'functional biology' – in which all these positions had their place, Woodger aimed to do nothing less than clarify and eventually solve all the biggest controversies affecting biological thought. On one hand, Woodger preferred a critical biology in the Kantian sense – a theoretical biology that, with the new powerful tools provided by the analytic philosophers, aspired to unravel the presuppositions, assumptions and dogmas preventing or hindering any form of agreement among opposing schools. On the other hand, and with the same analytic tools, Woodger tried to provide a logical and strong foundation to 'organismal' biology – a biology centred on the notion of organism as organized hierarchical system. In the next sections I will introduce Russell's biological thought and his context, and then I will focus on Woodger's background and his early theoretical biology. We will see that both were well connected, philosophically, socially and institutionally, with the scientific community of their time; in other words, they did not represent a small, silent minority but a rather noisy minority.

## E. S. Russell: The International Diffusion of Organismal Biology

> That we cannot explain or account for the directiveness and creativeness of life need make no difference to our projected functional biology; we must simply accept the immanent teleology of organic activities as, so to speak, the basis or background of our biological thinking ... because organic teleology is not mechanical, it is not therefore something miraculous and supernatural.[5]
>
> E. S. Russell

Russell's fame as biologist and philosopher went well beyond British borders; when he published *The Interpretation of Development and Heredity* in 1930, biologists in both German-speaking lands and the United States knew him and enthusiastically commented on his work. This is evident especially if we consider a letter that Ritter sent in 1931 from San Diego to a young Viennese philosopher who, after World War II, would become the initiator and promoter of 'general system theory', L. von Bertalanffy. Ritter had previously advised von Bertalanffy to read Russell's book; in turn, von Bertalanffy replied that he had already read it – indeed, Woodger himself had sent a copy to Austria:

> It is very nice on your part to bring to my attention Russell's 'Interpretation of Development'. I knew that book already, Dr. Woodger sent me a copy of it. While I was reading it, I was quite amazed by how similar the book is – in content and order – to my 'Critical Theory of Morphogenesis' (which was published almost 3 years ago). Since there is no reference in his work, Russell does not seem to know my publication; however, this is for me a remarkable proof of the 'publicity of the case' and, again, it proves how much our 'organismic' train of thought is in the air today. You will be further convinced of the striking parallelism between Russell's book and mine once it comes out in English, which, as you know, will happen very soon. In any case, I see the English edition of my 'Critical Theory', and especially my 'Theoretical Biology', as a formulation of the 'organismic' view that represents a synthesis of a large part of my stances expressed in the past and Russell's position on the matter.[6]

When the English edition of von Bertalanffy's book came out in England in 1933 (in a version translated by Woodger himself and titled *Modern Theories of Development*), he expressed, in the introduction, his enthusiasm about the fact that the organismic view in biology was shared among many biologists and philosophers in many different countries; it was in the air because great books and articles, he continued, had been written on such a subject, and figures such as Russell in the United Kingdom and Ritter in the Unites States led the way. Even a reviewer of von Bertalanffy's book considered his author as 'one of a small band of people who are paving the way to a new conception of the organism, a new orientation of biological thought'.[7]

Russell was born in 1887 in Port Glasgow, Scotland. He went to the University of Glasgow, where he studied Latin and Greek and only later started studying marine biology with the Scottish embryologist J. G. Kerr (1869–1957). His earliest works and interests were focused on the nature of coelenterates (in particular cephalopods).[8] During his years as a student, Russell's intellectual development was deeply indebted to J. A. Thomson (1861–1933). As we saw in Chapter 2, Thomson endorsed an anti-mechanistic, Bergsonian philosophy, a conception that he had also elaborated through the direct influence of Geddes. Thomson and Geddes thus exerted a deep influence on Russell. As M. Graham stresses in his short biographical sketch of Russell, it was Geddes who fostered Russell's interest in the philosophical aspects of biology. In 1921, under Thomson's guidance, Russell obtained the degree of DSc with a thesis titled, like his seminal book of 1916, *Form and Function.*[10] For the rest of his life, Russell remained an authority on marine biology and fishery investigations, publishing important articles on fishery statistics, fish behaviour and problems related to overfishing.[11]

Before World War II, Russell was not only an authority on marine biology but also a celebrated bio-philosopher.[12]Although his *Interpretation* further established him internationally, this cosmopolitan recognition was not new; he was already known in Italy, where in 1909 he had collaborated on a newly founded international journal directed by two Italian philosophers, Eugenio Rignano and Federigo Enriques. Indeed, in 1907 Rignano, Enriques and other fellows based in various Italian universities edited the first edition of *Rivista di Scienza* (renamed *Scientia* in 1909), aiming to offer an interdisciplinary platform where scientists from different countries, and interested in many different disciplines, could show and discuss their own specialities before an international audience. The editors defined the journal as an international organ of scientific synthesis, and in order to further such an international character, articles and reviews were presented in four languages: French, German, Italian and English. As well as von Bertalanffy himself, who, like Russell, worked regularly for the journal,[13] famous contributors included scientists and philosophers such as Freud, Neurath, Carnap, Driesch and Einstein, to name just a few.[14]

Between 1909 and 1933 Russell would publish two articles and more than 100 reviews in this journal. Moreover, he shared ideas, views and approaches in biology with Rignano. Above all, Russell was fascinated by Rignano's mnemonic theory of heredity. Rignano, who was seventeen years older than Russell, was born in Livorno, Tuscany in 1870. He had trained in physics and engineering at the University of Pisa and the University of Turin.[15] From his university years, Rignano had demonstrated deep interests in philosophical issues as related to science and politics. As Enriques recalled, Rignano used to read Spencer and Comte, as well as Mill and Kant.[16] With the new century, Rignano's philosophical and historical interests were increasingly focused on biology and its linked theoretical issues. Lamarckism, the opposition against neo-Darwinism, and holistic, anti-mechanist and anti-reductionist approaches characterized Rignano's bio-philosophy from the beginning. In 1926 he started a debate with Joseph Needham on the nature of life and organisms. The book he had published under Driesch's auspices, *Man Not a Machine: A Study of the Finalistic Aspects of Life* (1926),[17] triggered this harsh debate with the young Cambridge biochemist, who published an anti-Rignano pamphlet entitled *Man a Machine in Answer to a Romantical and Unscientific Treatise Written by Sig. Eugenio Rignano* the following year. Needham's stance was quite sophisticated. He was not a convinced mechanist, but he believed that mechanism was the best approach to undertake for any scientific investigation. In 1928 Needham published a short essay where his ideas were fully explicated. Organicist and holistic views were good and useful for philosophy, but they were useless in science. The scientist working in his lab needs to be a mechanist if he wants to progress in his researches: 'Neo-Mechanism would therefore seem to be the theory of biology best adapted for the future. For it only advocated mechanism as essentially methodologically, and did not claim that it gave a truer account of the world than the organicism of the philosopher'.[18]

However, Rignano's teleological and organismic views on biology met many admirers, especially on the continent. Von Bertalanffy was one of his allies and friends; he not only collaborated on *Scientia*, writing articles and reviews, in 1930 he also revised a new German edition of Rignano's book *Biological Memory*.[19] Rignano's obsession for mnemonic phenomena as associated with heredity and development dated back to 1906, when he published a moderately successful book entitled *Sur la Transmissibilité des Caractères Acquis: Hypothèse d'une Centro-épigénèse* – a work that, within five years, was translated to German, English and Italian, and that aimed to extend and develop the previous mnemonic theories of inheritance as advanced by people such as E. Haeckel, E. Hering, S. Butler, J. Ward, F. Darwin, J. A. Thomson and R. Semon. What Rignano tried to do in this book was twofold: to formulate a hypothesis that could synthesize preformationist and epigenesist theories of heredity, and at the same time to propose a Lamarckian mechanism through which acquired characters could be passed to the next generation. To Rignano, the element unifying ontogeny, phylogeny and the acquisition of novelty in evolution was a mnemonic mechanism; ontogeny recapitulated phylogeny because the fertilized egg retained, in the form of mnemonic traces, all the history of its ancestors. Just as a phonograph 'reads' a disc producing sounds that were previously recorded, development represented to Rignano a progressive unfolding of mnemonic traces 'recorded' during phylogeny.

Russell was quite enthusiastic about Rignano's attempt; in 1910 he wrote a paper for *Scientia* in which, following Rignano, he dismissed both preformationist and epigenesist theories of heredity and development, assuming that the only hypotheses logically coherent and in agreement with scientific data were the mnemonic theories which 'do not find many supporters among biologists nevertheless the compelling justifications that Semon, Rignano and Francis Darwin have provided long ago ... what it is really significant is that these theories virtually introduce a new method in biology, a method that can be recognized as properly biological, whereas the mechanist and morphologist methods preferred in our time are not biological at all'.[20]

The relationship between Rignano and Russell was further established when, in 1911, Russell travelled to Bologna in order to participate in the Fourth Congress of Philosophy; there Russell read a paper soon to be published on *Scientia*[21] – a paper aiming to reject both vitalism (and its new forms) and neo-mechanism and, at the same time, proposing a third 'attitude': an organismal approach. As Russell explained, 'One may concede the universal validity of physical and chemical laws and yet hold that the laws of biology cannot be reduced to their level. To admit the physico-chemical determinism of life is not to admit that physico-chemical laws are adequate to explain life'.[22] Like Haldane before him, Russell believed that life presented and exhibited peculiar phenomena, phenomena necessarily irreducible to physico-chemical mechanisms: 'while many of the phenomena presented by living things are thus to be explained as the direct result of simple physical and chemical relations, there still remain a vast number of facts of life which cannot be explained by any direct reference to chemical laws. They present truly biological problems which can be solved only by biological laws'.[23] One of the central phenomena that Russell deemed unexplainable in physico-chemical terms was the migration of eels. In 1909 Russell had joined the Board of Agriculture and Fisheries; working at the Aberdeen Laboratory, his research involved, among other things, studying the effects of intensive fishing. In particular, the fisheries department was committed to formulating censuses based on statistical observations of fish populations in various areas of the ocean. The fluctuations of fish populations, their distribution and their migration in relation to the environment (without overlooking food chains and behaviour) provided precious information for a rational exploitation of the seas' resources – vital information for the national economy.[24] As a skilled observer of marine biological phenomena, Russell maintained that knowledge about all these issues had to remain on a purely biological basis.

To Russell, the fact that eels migrated during the breeding season from Northern and Western Europe to the warmer, saltier and deeper waters of the Atlantic represented an irreducible living phenomenon because:

> even if we knew the exact chemical mechanism of muscular contraction, and of nervous conduction, for muscular movement is dependent upon nervous stimuli, we should not be a whit nearer an explanation of the fact that eel was taking this long journey to a particular area of the North Atlantic for the purpose of spawning. The fact is that the laws of metabolism and of the physico-chemical side of life lose by their very generality all power to explain concrete facts of higher order.[25]

In short, migration 'emerged' from all the chemical and physical elements put together in forming a higher order of phenomena. As Russell concluded:

> In the case of the eel it is possible to decompose the act of migration into a large number of acts of a different order, into the chemical reactions occurring in muscular movement, in nervous conduction, in the stimulation of peripheral sense organs, but by doing so one cannot but lose sight of the interconnection of these single acts, the interconnection which really binds together all these acts into the single act of migration.[26]

At the Bologna conference Russell concluded that biology was an independent and irreducible science, not because living organization required entelechies or vital principles but because it dealt with phenomena of a peculiar order. In the following years, in reviewing and criticizing more than 100 books in the pages of *Scientia*, Russell would state such a position over and over again. His vitriolic reviews against Mendelian genetics and all books propagandizing a mechanistic view of life (including, in his opinion, Bateson's, Morgan's and Loeb's books) established him as an organismic biologist. There are very few doubts that Russell's publications in *Scientia* contributed in no small part to his international fame.

However, in the next sections we will see that Russell's plan throughout the years – from the publication of his first book (*Form and Function*) to his last contribution (*The Directiveness of Organic Activities*) – followed a very coherent and reasoned line of thought. Both history and philosophy had to support his conception of biology, a conception that, as we will see, was based on functional and integrated organic activity. Yet Russell's theoretical biology was employed concretely to deal with many different issues, from the interpretation of animal behaviour to regeneration, and from the physiology of digestion to comparative anatomy. But especially, it was used to tackle the classic problems of biology since Aristotle: development and heredity.

## Russell's Agenda: Between Science and History

Although Russell would eventually be committed throughout his life to the study of marine biology as related to fisheries issues – becoming the first Director of Fisheries Investigation at the Ministry of Agriculture and Fisheries and, in 1940, president of the zoology section at the British Association and president of the Linnaean Society – today he is mainly remembered for one of his classic books: *Form and Function: A Contribution to the History of Morphology*.

The book was published in 1916, when he was twenty-nine, and represented a general excursus through the history of biology – a history that, nevertheless, had to support Russell's biological agenda. In fact, more than anyone else I have mentioned before or will discuss later, Russell was fully aware of being part of a venerable tradition: the post-Kantian, Romantic tradition. As a keen historian, he selected his sources wisely and formulated a coherent narrative supporting and legitimizing his own preferred option: the triumph of functional biology over materialism, the so-called 'aberration' of the nineteenth century. Indeed, Russell's conclusions reflected his social concerns; to him scientific materialism was not the result of a wrong philosophy but the outcome of the 'over-rapid development of a materialistic and luxurious civilization, in which man's material means have outrun his mental and moral growth'.[27] Russell shared some of the preoccupations of the French neo-Lamarckians. With Alfred Giard (1846–1908), a zoologist whom Russell knew and quoted in *Form and Function*, he felt that the association between neo-Darwinism and materialism supported a 'heartless capitalism'.[28] Materialism was not only wrong as a scientific interpretation, it was a dangerous ideology to be fought against: 'if social reform were ever to be achieved', Persell reminded about the neo-Lamarckians, 'it had to come in a flexible milieu where environment could be manipulated and the individual changed'.[29] History and science were the tools with which Russell chose to fight the battle against materialism, determinism and their more diffused instantiations: neo-Darwinism, Weismannism and Mendelism.

The main argument characterizing and ordering Russell's book was the idea that the history of morphological sciences (or morphological thought) could be divided into three different schools: the functional or synthetic school (to which Aristotle, Cuvier and von Baer belonged); the formal or transcendental school (of which Geoffroy Saint-Hilaire was a representative); and finally the materialist or 'disintegrative' school (which was founded by the Greek atomists). In Russell's passionate narrative, Cuvier emerged as a hero: after Aristotle he was the first truly comparative anatomist because, as Russell claimed, 'like Aristotle, like the Italian anatomists, Cuvier studied structure and function together, even gave function the primacy'.[30] Cuvier, he added, was a teleologist in the wake of Kantian bio-philosophy; indeed, as Russell explicitly reported, 'there can be no doubt that [Cuvier] was influenced, at least in the exposition of his ideas, by Kant's *Kritik der Urteilskraft*, which appeared ten years before the publication of the *Lésons D'Anatomie Comparée*. Teleology in Kant's sense is and will always be a necessary postulate of biology'.[31]

Even though Cuvier's principles of correlation and condition of existence represented, to Russell, two fundamental pillars supporting what he would eventually dub 'functional biology' or the 'psychobiological approach', he thought that other figures had made central contributions: Goethe and the German transcendentalists, and Richard Owen in Britain.[32] However, another giant surfaced in the history of biological thought: K. Ernst von Baer: 'Von Baer reminds one greatly of Cuvier. There is the same sheer intellectual power, the same sanity of mind, the same synthetic grip. Von Baer, like Cuvier, never forgot that he was working with living things; he was saturated, like Cuvier with the sense of their functional adaptedness'.[33] According to Russell, von Baer was influential well beyond his own country. In fact, not only was he behind Darwin's phylogenetic tree because 'Darwin interpreted von Baer's law phylogenetically',[34] but he also influenced different generations of scholars who, like J. Müller, disapproved of cell theory and its atomist interpretations and, like T. Huxley, defended the notion of ideal type.

In 1911, at the conference in Bologna, Russell had stated that living organisms were different from inert matter because, among other things, they were historical entities; activities, forms and structures were necessarily shaped and related to their experiences and, therefore, any living being could only be understood comprehensively by knowing its phylogenetic past.[35] Now, five years later, Russell was arguing that knowing organisms' past would involve the understanding of hereditary transmission; indeed, knowledge about heredity would help to distinguish which characters were acquired and which simply inherited during evolutionary history. However, with hereditary transmission, Russell did not intend gene transmission or any particulate theory of heredity; he meant a functional science of heredity linking together embryology and morphology. As a Lamarckian, Russell was convinced that functional activities of the organism slowly changed its structures, while structural change could be fixed and transmitted to the next generations. Indeed, the enormous appeal that mnemonic theories of heredity had for Russell, and that closed his celebrated and ambitious first book, derived from his Lamarckism. Because structural variations depended on the organism's functional responses to the environment and functional satisfaction of its needs, structural changes emerged from those functional activities that, in some way and under different forms, were 'stored' in the germ cells and became hereditary. Samuel Butler, together with Orr, Cope, Rignano and Francis Darwin, were enlisted in supporting such a view (although from slightly different perspectives).

Russell's biological agenda – an agenda now justified from his historical account – found further support in his theoretical biology (and its philosophical foundations). Indeed, in 1924 Russell published a new book simply titled *The Study of Living Things*, a work aiming to 'outline' a method for biological investigation that would eventually avoid both materialist approaches and vitalist speculation. Haldane's influence on Russell's second book was particularly evident; in fact, as we have seen in the previous chapter, during the second decade of the twentieth century Haldane became a prominent figure among scientists, philosophers and laymen. Haldane symbolized a campaign against mechanist and reductionist interpretations of physiology and biology, and Russell did not hide this; as he clearly pointed out in the book's preface: 'I owe much to the writings of J. S. Haldane'.[36] This was not, however, a mere vague acknowledgement, because Haldane's ideas were effectively reported and positively discussed throughout Russell's book; indeed, to him, Haldane's work represented the ideal method to follow in biological investigations:

> The finest modern exposition of a 'biological' method, which should mediate between physics and psychology, is that set forth by Dr. J. S. Haldane in a series of recent publications. In considering his powerful plea for an independent biology with laws and concepts of its own we shall see how irreconcilable are his views with the mechanistic theory[37]

In his epistemic treatment of life sciences, Russell distinguished five different methods that had been applied and used in different historical periods. There was a morphological, a physiological, a vitalist, a psychological and a biological method. The first assumed that an organism's activities could be deduced from observation and knowledge of its structure, so that function was an 'effect' of the organism's material structure. Yet Haldane in England and Delage in France had convincingly rejected such a method: it betrayed an antiquated 'habit of thought', it supported the misleading machine analogy, it sustained a static view of living entities, and it backed particulate theories of heredity. The second 'attitude', although recognizing the primacy of function upon structure, overlooked the organism's unity, its individuality and its 'relative independence' from the environment:

> Materialist physiology ... goes to the opposite extreme from morphology. While the morphologist studies the form and structure of the organism in almost complete isolation from its environment, the physiologist merges it completely into its surroundings and robs it of all independence ... it cannot consider the organism as a living and independent whole. Its laws have no reference to individuality.[38]

The third method was quickly dismissed; indeed, vitalism was defined by Russell as pure materialism with the addition of a mystical, speculative and unintelligible force. The fourth viewpoint represented a positive step towards the 'right' method, as it assumed the individuality and unity of organisms and presupposed their abilities to perceive the environment and behave accordingly. Finally, the fifth approach represented the best so far – it was Haldane's approach. It regarded organisms as persisting, real and individual wholes.

However, Russell advocated a mixed method: both psychological and biological approaches could be used together to make a psychobiological method, which was dubbed simply 'functional biology'. Russell had probably borrowed the word 'psychobiology' from Bergson and the German botanist R. H Francé. Dissatisfied with scientific materialism, Francé thought that life could not be explained by physics and chemistry. Influenced by Kant, Nietzsche and Schopenhauer, in 1921 he published a monumental work, *Bios. Die Gesetze der Welt*

(Bios: The Laws of the World), that attracted the attention of Geddes, D'Arcy Thompson and, of course, Russell.[39] The psychobiological approach had to assume not only that organisms were perceptive systems but also that they were irreducible purposive individualities. Living things always have 'instinctive tendencies' to keep such a system intact; they strive, struggle and fight to stay alive. Such a set of instinctive tendencies was named by Russell 'hormé':

> The objective purposiveness of organic activity, which is the outcome of *hormic* impulse, or tendency, in not, like the purposiveness of a machine, a fixed and automatic thing, but on the contrary is self-regulatory and adjustable to a wide range of circumstance. A completely blind, unalterable and automatic activity would soon destroy itself ... adjustability, flexibility, of response is essential.[40]

Therefore both biological and psychological approaches were strictly related because organisms, according to Russell, acquired their specific individuality and structure through active perception of their environment and, accordingly, through their responses and adjustments to it.[41] The characterization of organisms as '*hormic* interconnected systems' was beautifully described by Russell with a very Kantian analogy between works of art and natural products (Kant's *Naturzwecke*):

> There is an unanalysable unity in a work of art – the parts form an individualized whole or unity and have a meaning only in relation to the whole. So also in the living thing. As works of art are static organisms, so organisms are dynamic works of art. It follows that living things can, no more than works of art, be exhausted of their content by the analytic and superficial description offered by the physical sciences. They are the sounds and the shapes and the colours, but they are more, as works of art are more. They differ from works of art being self-creative, or created by Nature as Artist.[42]

The functional method advocated by Russell was readily applied to specific biological problems: first of all, to the interpretation of the behaviour of the simplest organisms. In contrast to Loeb's rigid and mechanist interpretations, Russell noticed that organisms such as *Amoeba* or *Pelomyxa* did not show very stereotypical or fixed behaviour; in the actions involved in capturing motile prey, they manifested creative and unexpected activities. If organisms reacted in unexpected ways to old or new situations (old or new stimuli),[43] a mechanical interpretation based on automatic or determined responses had to be rejected.[44] With Bergson – who in his *Creative Evolution* had argued that *Amoeba*'s movements could not be understood through physics and chemistry[45] – Russell believed that both *Amoeba* and *Pelomyxa* responded not to a physico-chemical stimulus but to a perceived and interpreted stimulus. In other words, the best way to interpret creative behaviour is to assume that external stimuli were 'perceived' as directly 'significant', and not as mere external physico-chemical stimuli: 'What is responded to is not the stimulus *qua* physico-chemical, but the stimulus as perceived, and not the stimulus merely as perceived, but as interpreted. Response is really to the meaning of the perceived stimulus, not to the stimulus itself'.[46]

## Negative and Positive Conditions of Life

In addition to Haldane, Russell also enrolled D'Arcy Thompson in his broad synthesis. They shared ideas and interests, and Thompson admired Russell's book *Form and Function*. As he told Russell in private: 'I have been reading your new book very carefully and with great pleasure, and I should like to send you my hearty congratulations and my best wishes for the book's success and your own'.[47] In turn Russell praised Thompson's method in his *On Growth and Form*:

> the important thing, from my point of view, is that you bring out the universal existence of correlation in Cuvier's sense. I tried to point out in my book the great value of the Cuvierian conception, which has for long been forgotten and ignored. It would be interesting to treat cases of functional convergence by your method as well as cases of homological resemblance. And the method should be extremely valuable applied to growth-changes.[48]

In Russell's view, although functional biology entailed a study of 'perceived' stimuli and functional adaptations, it did not overlook the so-called 'material condition of life', that is to say, all the properties of life that are contiguous with physics and chemistry. Because functional activities required specific substances to work and were constrained by determined physical forces, knowledge about what Russell dubbed 'the negative condition of functional activity' was essential, and Thompson's *On Growth and Form* pointed in that direction. Of course Russell deemed physico-chemical investigation quite a secondary interest for biologists; after all, physico-chemical processes, elements such as temperature, various chemical substances essential to life, light, etc. represented a mere background within which organisms actively moved and against which they creatively reacted. Like Haldane and Thompson before him, however, Russell maintained that investigation in biology required both analysis and synthesis; therefore, knowledge of the organic parts constituting the whole (physico-chemical as psychobiological knowledge) and how the whole produced or shaped its parts was required. The physico-chemical understanding of the organism represented only a small step towards a comprehensive understanding of the nature of life.

Even though animal behaviour and the so-called 'negative conditions of living' were both important parts of Russell's functional biology, one of the most significant theoretical consequences of functional biology was its incompatibility with Weismann's germ plasm theory and with Mendelian genetics. For Russell, all of these theories (i.e. hypotheses based on the assumption that some well-defined particles could explain, in principle, the transmission and development of morphological characters) were based on pure speculation or presupposition. Russell did not question the idea that there was a material continuity among generations or races; he only questioned that such a continuity could be explained by determinants or genes. Functional biology must interpret both heredity and development from a very different perspective. First, heredity did not consist in the transmission of material stuff but in the transmission of tendencies (namely, potential functions); second, and as a consequence, development was the process through which these tendencies or potential functions became, gradually, actual functions: 'What is essentially transmitted is not structure but potential function, a bundle of habitual tendencies which gradually become actualised in structure'.[49] The link between functional or organismal biology and its bearing on the interpretation of heredity and development assumed a central position among his interests. After all, Russell's synthesis was precisely framed to tackle all the fundamental activities concerning living organisms, i.e. development, heredity, physiology and evolution. Indeed, in 1930 Russell published an entire monograph focused on the interpretation of heredity, development and their mutual relations. From a historical viewpoint, such a monograph probably represents one of the most vigorously anti-Weismannian and anti-Mendelian books written in that period, on British soil at least.

## Heredity and Development without Genes

> Delage's criticism of particulate theories is in fact unanswerable. If the cell and the developing organism are regarded as physico-chemical systems in constant metabolic relations with their environment, external and internal, there is absolutely no place for independent and isolated material units which represent or determine certain characters or groups of characters. No part of the cell can exist in isolation from the whole; to imagine such is to create a conceptual fiction to which nothing corresponds in reality.[50]
>
> E. S. Russell

For Russell, *The Interpretation of Development and Heredity* represented the consequent and most natural application of the organismal or functional biology he had previously established, both historically and philosophically, in previous books and articles. For many years he had expressed his ideas on these topics in reviews of French, German and English books in *Scientia*. In 1914, for example, he reviewed both Bateson's *Mendel's Principles of Heredity* and Plate's *Vererbungslehre mit besonderer Berücksichtigung des Menschen* (Heredity, with Special Reference to Man). Bateson's book was published the previous year and represented the third edition of a successful textbook. Russell praised the clear exposition of Mendelian doctrine, he liked the illustrations and pictures, and he appreciated both Mendel's biography and the translation of two of his papers. However, a few lines later Russell was much less enthusiastic; while he partly accepted Bateson's suggestion according to which hereditary units could be thought of as things similar to enzymes, he warned that Mendelian analysis was going too far:

> For now, Mendelians manifest a certain overconfidence, too much enthusiasm, about the universal application of their ideas ... Mendelism should regard the only appropriate place convenient to itself, which is that of a very precious method of analysis applicable to some modality of hereditary transmission; it could not be considered as a complete solution to all problems of heredity.[51]

In the next review Russell was less diplomatic; indeed, he complained that 'The conception of the organism as a set of character-units is dogmatically seen as the only possible alternative, and the study of heredity is arbitrarily defined as the study of gene's mutual relations (or determinant of unit-characters)'.[52] In the following years Russell would remain loyal to this idea. Reviews of T. H. Morgan, M. Caullery, L. Doncaster, H. F. Osborn and others betray his uneasiness with a gene-centric or unit-centric approach to many problems of biology, from sexual determination to human intelligence, development to evolution, morphology to adaptation. Mendelian genetics had to be considered only an aspect, a portion, a tiny section of a bigger discipline or larger framework exemplified from the studies of heredity, development and evolution. When *Interpretation* was published in 1930, Russell intended to reinforce this idea: heredity cannot be defined as Mendelian genetics because heredity was much more than that; indeed, heredity deals not only with stochastic variations among populations but also with the reproduction of individual organic form through development:

> The broad fact of repetition of type has tended of recent years to become lost from sight, because of the excessive attention paid to the laws of transmission of such slight variations or differences as are no bar to successful interbreeding; the study of heredity has come to mean in practice the study of the modes of inheritance of minor differences. But clearly there is this major problem which is practically untouched by genetic or statistical studies, and equally clearly, repetition of type must be regarded as one of the main characteristics of development, not as a separate and independent problem.[53]

In the same year that *Interpretation* was published, Fisher's *Genetical Theory of Natural Selection* appeared in print. Fisher's book propounded a very opposite conception of heredity, one totally based on populations rather than individuals, on discrete genes rather than repetition of types. Fisher believed that qualitative individual changes could be ascribed to the selection of small, measurable and quantitative variations within populations: variations inherited and selected. Neither evolutionary change nor heredity had anything to do with individuals and their development but instead with the transmission of genes as statistically measured within populations. Organic form passed second-hand; it became a mere epiphenomenon of gene-pool variations within populations. Such a doctrine was eventually to lead to the establishment of a new discipline called population genetics, the cornerstone of the modern evolutionary synthesis.

Russell, however, had a very different modern synthesis in mind, one based on his conception of functional or organismic heredity as the expression of developmental forces; in fact, the whole of *Interpretation* presents a long and critical argument against Morgan's idea that heredity itself could be severed from development. Aristotle, Russell argued, was the first to recognize 'the important point that the explanation of hereditary resemblances is dependent upon the explanation of development, that this resemblance is a feature of development rather than a separate problem'.[54] In Russell's interpretation, von Baer's whole notion of '*Wesenheit*', the essential nature of the animal, spoke against the idea that embryogenesis could be disconnected from hereditary processes, in principle at least. In fact, for von Baer, reproduction was defined as 'the taking on by a part of the potentialities of the whole', i.e. an organic part, a bud or egg, broke the domination of the whole and developed into a new individual organism.[55] The formation and development of this new individual was driven by the *Wesenheit*, the *essential nature*, which guided the organism from a fertilized egg towards its *complexification* and increasing individuality. The whole process was conceived as intrinsically developmental throughout: 'it is not the matter, in its mere arrangement, but the essential nature of the procreating organism that rules the development of the offspring'.[56] We could interpret the *Wesenheit* as the variable *x* that organisms transmit to their offspring, generation on generation; however, this hypothetical *x* could never be severed from development because development was the actual and explicit manifestation of *Wesenheit*. As a consequence, Russell thought that development explained evolutionary diversification. In fact, organic novelties were the result of developmental 'alterations', systemic changes that could be stored, or engraved, in the form of mnemonic traces in the germ cells and passed on to the following generations. Ontogeny, through the mediation of the environment, created phylogeny; or, to put it another way, development provided the raw material for new functional adaptations in evolution. Russell unified and mixed Lamarckism with *Naturphilosophie*, mnemonic theory with developmental biology, physiology and morphology – all within an organicist, anti-reductionist and functionalist framework.

## Against Weismann and Morgan

> The gene is a word, which enables a complicated happening to be briefly denominated.[57]
>
> E. S. Russell

Russell, in common with many organismal biologists, considered both Weismann's germ plasm theory and Morgan's theory of the gene as lying on a similar conceptual framework; in fact the notions of determinant and of gene had some interesting similarities, although their existence was deduced in different ways.[58] In 1926 T. H. Morgan had published *The Theory of the Gene*, a book in which he synthesized many of the results achieved to date by him and the group of students working at the Columbia Lab. Morgan summarized his theory succinctly: the exposition of Mendelian laws of segregation and dominance was followed by the hypothesis that genes were physically aligned along chromosomes. The phenomena of crossing over, linkage and mutation were explained and illustrated, the statistical, quantitative and physical studies on gene activity highlighted and praised. The book was certainly one of the most popular formulations of the gene theory so far and, for this reason, a central target for Russell.

To begin, Russell compared Johannsen's conceptions of genetics, genes and heredity with Morgan's. Wilhelm Johannsen (1857–1927) was a Danish botanist who first coined the term 'gene'. In 1923 he published a very short article that Russell took as a source of inexhaustible inspiration. The article was published in *Hereditas*, a journal that was originally edited by Mendelska sällskapet i Lund (the Mendelian Society of Lund), and it represented a formidable attack against the most radical Mendelian agendas. Firstly, Johannsen was sceptical about ontological interpretations of genes as material or physical entities aligned on chromosomes stored in cell nuclei.[59] Secondly, his notion of genes betrayed a developmental conception; to Johannsen, genes were not things but correlated chemical *reactions*, whole organic reactions that had nothing do to with the notion of the unit-character: 'from a physiological or chemico-biological standpoint we must *a priori* in characters or developed parts of organisms see *Reactions* of the (I should say geno-typical) constitution belonging to the zygote in question; and from this point of view there are no unit-characters at all!'[60] Thirdly, genetics was affected by misleading terminology guiding a confusing train of thought: 'in the language of genetics we meet with some unhappy old-fashioned expressions, relics and obsolete conceptions – the worst of all these relics is probably the expression *Transmission* where no transmission exists but where continuity is found!'[61] Fourthly, the hope of reducing the genotype to a set of particles was pretentious and unsupported by observed facts; as Johannsen himself emphatically put it:

> We are very far from the ideal of enthusiastic Mendelians, viz. The possibility of dissolving genotypes into relatively small units, be they called genes, allelomorphes, factors or something else. Personally I believe in a great central 'something' as yet not divisible into separate factors. The pomace-flies in Morgan's splendid experiments continue to be pomace-flies even if they lose all 'good' genes necessary for a normal fly-life, or if they be possessed with all the 'bad' genes, detrimental to the welfare of this little friend of the geneticists. Disregarding this (perhaps only provisional?) central something we should consider the numerous genes, which have been segregated, combined or linked in our modern genetic work. But what have we really seen? The answer is easily given: we have only seen *differences*.[62]

The very same distinction between genotype and phenotype that Johannsen first made was essentially developmental; to him, in fact, the phenotype represented the reaction of the whole genotype in interaction with the environment: 'however far we may proceed in analysing the genotypes in separable genes or factors, it must always be borne in mind, that the characters of the organisms – their Phenotypical features – are the reaction of the genotype *in toto*. The Mendelian units as such, taken *per se* are powerless'.[63]

To Russell, all this closely fitted his own conceptions: Johannsen's central 'something' could easily be interpreted as von Baer's *Wesenheit*; heredity was 'something' that, in principle, could not be cut and sliced because it was a property of the whole organism. The gene theory instead reified and endowed 'with material existence what are merely differences, and it does this by postulating a gene for every heritable difference found'.[64] Morgan, together with Mendelian geneticists, invented or posited fictional entities to explain the differences observed in *Drosophila* eye colour. In so doing, they sliced the organism into abstract parts: on the one hand heredity, with its fictional, discrete and separate particles; on the other development, conceived as mere effects of those particles and their arrangements. However, not only was Mendelian genetics inadequate as an explanation for the mechanism of heredity in general; it also focused on unimportant or secondary phenomena.

As Albert Brachet (1869–1930), one of the leading figures in the Belgian school of embryology, had famously distinguished, development dealt with two different kinds of hereditary potencies or tendencies: those characterizing *general heredity* and those making *special heredity*. The former has to do with the laws or causes producing the whole individual belonging to a species – the general type, so to speak. The latter instead concerned small individual variations – or, in other words, all the variations or deviations added to the *general heredity*.[65] In Brachet's as in Russell's mind, special heredity represented a secondary subsection of general heredity, which denoted the study of heredity *par excellence*. Now genetics, in dealing with small or superficial variations within special heredity, overlooked the production and reproduction of the organism's type, i.e. general heredity. In other words, genetics was unable to shed any light on one of the oldest issues in biology: how like begets like.[66] Furthermore, drawing on F. R. Lillie (see Chapter 5),[67] Russell added that gene theory, as it was formulated, was of no help in understanding how organisms develop. Positing imaginary entities equally stored in the cell nuclei did not advance our knowledge of the way that cells differentiate; indeed, differentiation was probably a phenomenon tied to cell cytoplasm and internal cell interactions in connection with the whole individual organism. In the end, Morgan's gene theory was considered by Russell as inferior even to Weismann's germ plasm.

Russell's project to propose a form of heredity and development devoid of particles included both historical figures and contemporary scientists: Aristotle, von Baer, Haldane, Johannsen and Lillie of course, but also C. Bernard, Y. Delage, D'Arcy Thompson, C. O. Whitman, C. M. Child, C. Sedgwick, C. Sherrington, E. G. Conklin, H. Woodger, W. E. Ritter and others.[68] Ritter was particularly important, as Russell had borrowed from him the term 'organismal biology'; as Russell stated: 'It will be seen that Ritter's point of view is essentially the same as that taken here, though he has arrived at it by a different route. I take this opportunity of acknowledging my indebtedness to Ritter'.[69] Bernard clearly upheld and disseminated the organismic view in medicine and biology.[70] Delage too deserved a special mention in Russell's intellectual geography, as in 1895 he had published a large volume, *L'hérédité et les grands problèmes de la biologie générale*, a book that went through three editions and that influenced subsequent generations of scholars. Delage had introduced the term 'organismic biology', had formulated a violent critique against particulate theories of heredity, and had proposed an epigenesist hypothesis of heredity and development – all achievements that Russell found highly significant. D'Arcy Thompson had successfully criticized the morphological, structural approach to understanding heredity and development; to Thompson, heredity (as we have seen in the previous chapter) had to do with energy and its manifestation and not with bare matter. Yet to Thompson, heredity was a property belonging to the nucleus, cytoplasm and the cell as a whole. Furthermore, Thompson had also conceived organisms in a Kantian way: as individual wholes where the sum was more than the parts.[71] Finally, with Whitman and Conklin, Thompson had condemned cell theory: the individual organism was not a mere colony of cells because cell organization was itself a product of the whole organism. Furthermore, Child had clearly shown that ontogeny was triggered and directed by a whole system of reactions; neither particles nor vital entities were required. The organism was seen as an indissoluble unity in both heredity and development.[72] Both Sedgwick and Sherrington had stressed the functional integrity of the organism as a whole, the former in his embryological observations and the latter in his studies of the nervous system. Finally, Woodger had analysed the meaning of some central notions belonging to organismal biology.[73]

Russell was a synthesizer; he gathered information from many different sources, discussed the pros and cons of several theories, and built his own coherent system of thought – a system based mainly on a strenuous critique of mechanist, materialist and reductionist interpretations of living phenomena. Like Haldane, he believed that scientific materialism had dangerous social consequences, as Nils Roll-Hansen explains:

> He [Russell] sees the refutation of reductionism in biology as part of a crusade against materialism. But he believes mechanism to be a failure in biology, as well as philosophically untenable and socially harmful. Mechanism had dominated nineteenth-century biology, and now it was time to formulate a fundamentally different methodology.[74]

However, Russell's theoretical system – his organismal discourse – was not entirely original. Indeed, as von Bertalanffy had mentioned in his *Modern Theories of Development*, organismic or organismal biology was a fairly widespread movement, both in Europe and in the USA:

> If one may use a pictorial catchword the development of science in the last decades may be characterized by an 'organismic revolution' which has taken place ... the reader on this side of the Atlantic will easily detect the parallelism with Whitehead and other American authors who, at this time, were very little known in Europe. This 'revolution', though, was not merely a development of philosophy but most definitely in science itself.[75]

The 'organismic revolution' mentioned by von Bertalanffy had its origins, in fact, in a venerable and long tradition that increasingly became philosophically convincing and internationally recognized. The application of such a philosophy to the problems of heredity and development provided radical and controversial results. The tenets of the organicist tradition clashed with so-called 'modern' hypotheses of heredity and development; as Russell explicitly concluded, genetics, together with all particulate hypotheses, had to be rejected, first and foremost, on theoretical grounds:

> If we hold fast to the principle that the whole cannot be completely explained in terms of its parts, that the modes of action of higher unities may be conditioned, but cannot be fully accounted for, by the modes of action of lower unities, it follows that no substance and no sub-cellular unities can be invoked as sufficiently accounting for the phenomena of development and heredity, which are essentially phenomena manifested by whole organisms – unicellular or multicellular.[76]

For Russell, chromosomes had to be seen as starters of catalytic processes in an integrated hierarchical system; they did not cause development and differentiation on their own; they were not alone responsible for hereditary characters. Indeed, if the higher activities of the organism were not fully reducible to its lower actions, the activities of the whole could only be transmitted by the whole itself, namely by the whole fertilized egg. Heredity had to be seen as the *actual* functional exhibition of *potential* tendencies, tendencies that could be identified only through the study of development. Even after the triumph of the chromosome theory of heredity, Russell would continue to support mnemonic, Lamarckian and developmental interpretations, which were increasingly enriched by the strong belief that organic phenomena were understandable as hierarchical – partially reducible and essentially interrelated – phenomena. Yet he continued to maintain that evolutionary diversification was not the result of gene mutations and selection, but was the outcome of individual developmental 'alterations', complex modifications transmitted – or engraved – in the germ cells and inherited. In short, whereas heredity had to do with mnemonic systems that controlled ontogeny, evolution resulted from the changes of mnemonic systems owed to repeated functional stimuli and environmental stresses. Russell synthesized Lamarckism and *Naturphilosophie*, mnemonic theory and developmental biology (including physiology, morphology and ethology), all within an organismal and functionalist paradigm. However, it would be Woodger who, more than anyone else, would dedicate – with the relatively new logical tools provided by analytic philosophers – the greatest attention to formulating a coherent and well-grounded model of the organic hierarchical system.

## J. H. Woodger: What Biology Required

Russell had argued that the difference between the organic and inorganic world reposed on a direct, evident and obvious perception derived from common sense.[77] 'Uncritical common sense', argued Russell, 'distinguishes sharply between living things and lifeless things'; yet human common sense could so easily distinguish animals from stones or clouds because organisms 'possess a power of individualized activity. The living thing is active and individual, the lifeless thing is passive and unindividualized'.[78] To Russell, it was common-sense perception of living phenomena that guaranteed, *prima facie*, the independence of biology from other sciences; biology was different and had its own 'rules' and 'laws' because organisms were *evidently* different, ontologically and epistemically diverse, from stones or clouds. The philosophy of common sense, based on a simplified and naive empiricism, was also part of Woodger's theoretical baggage with respect to the epistemology of biology; his recurrent quotations of Bacon, Hume or Locke clearly demonstrate his fascination for empirical philosophies and common-sense attitudes. However, although Woodger maintained that experience – as characterized through careful observations and detailed experiments – was an essential part of the scientific enterprise, he also held that a critical and theoretical analysis was required in any scientific investigation, including biology. With Whitehead – who had stated that science had to become philosophical and critical of its own foundations[79] – Woodger believed that zoology was besieged from a 'multitude of facts' without theoretical order. As he introduced his major philosophical work, *Biological Principles*:

> Modern natural science may be likened unto a crab which has grown too fat for its shell. The process of ecdysis is slow and painful. The old shell, which has maturated and hardened for some three hindered years, has done good service. No wonder the crab is loath to part with it. But it has already begun to crack, and some bits have even dropped off. What is to be done?[80]

In the crab analogy, the internal matter, the meat, represented the data – all the evidence scientists had gathered so far – whereas the shell symbolized the philosophical framework ordering and systematizing that data into a coherent whole. To Woodger, the philosophical background was collapsing under the relentless and disorganized accumulation of data. What was to be done? Woodger believed that a possible solution lay in a new philosophy, one able to match a Kantian critical approach with the new tools offered by analytic philosophy. Indeed, what biology required was not more evidence, but more thought – more criticism and more philosophy. A Kant scholar, Ernst Cassirer, had shown how important theories of knowledge had been in the history of physics linked to experimental analyses; methodology and experimental investigations went hand in hand in Galileo's, Kepler's and Newton's scientific revolution.[81] But while in physics theoretical speculations were considered part of scientific investigation, in biology philosophical treatments were dismissed, criticized or, at least, regarded as suspect. Woodger aimed to reverse that situation and to show that theoretical biology was essential in solving issues that remained unsolved, even with the help of further planned experiments and specific observations. Many biological issues, indeed, rested on conceptual ambiguity and theoretical misconceptions. In Woodger's opinion, the philosopher of biology, before choosing one theory rather than another, one approach over other methods, had to investigate all assumptions, presuppositions and postulates supporting those hypotheses or methods.[82]

The philosophical method Woodger employed was essentially Kantian:[83] identify a controversy between two factions and undertake an analytic investigation into the assumptions that each party accepted. In Woodger's hands, Kantian critical philosophy became the analytic investigation of intellectual controversies: structuralists against functionalists, preformationists against epigenesists, vitalists against mechanists, organism against environment, and inherited against acquired characters – all debates that Woodger analysed by dismantling, through a detailed *exegesis* of the arguments involved, the hidden postulates making disagreement possible. Of course, Woodger's theoretical philosophy was much more than that; indeed, as I will show in the next sections, Woodger provided an analytic and solid foundation for organismal biology.

## Analytic Biology: From Practice to Theory

Roll-Hansen defined Woodger as 'a logical empiricist, drawing inspiration from Kant's philosophy of science and from linguistic analysis'.[84] I would slightly change the order of Roll-Hansen's sentence and define him as a Kantian at heart, 'drawing inspiration from logical empiricists and from linguistic analysis'. This is not a matter of trivial detail; it makes a substantial difference to our interpretation of Woodger's bio-philosophy. Indeed, logical empiricism, however defined, was only one aspect of the complex and heterogeneous philosophy of the young Woodger. Of course, in the late 1940s his interests increasingly narrowed towards a classic logical positivism *à la* Carnap. In fact, after World War II, Woodger attempted to provide an effective, though vain, axiomatization of biological thought.[85] However, when we consider Woodger's overall earlier contributions, we get a quite different and more complicated picture. With Woodger there were two related agendas: a *critical* bio-philosophy aiming to clarify terms, notions and unsolved issues, and a *positive* bio-philosophy aiming to support and clarify the notion of organism as an integrated and teleological hierarchical system.

Woodger's philosophical insights had few followers; like Haldane, D'Arcy Thompson and Russell before him, he founded no school or tradition continuing his work, either in philosophy or biology.[86] Furthermore, Woodger's philosophy had no impact on the modern synthesis movement and its advocates; this not only because we have no evidence attesting the link between Woodger and any of the actors of modern synthesis,[87] but also because, as J. Cain claimed:

> his project to 'unify' biology also was an enterprise *different in kind* from the intellectual projects long identified with synthesis actors. His criteria and methods for unification were fundamentally different from what counted as 'synthesis' among evolutionists. Their epistemic concerns sharply contrasted. There is no evidence to suggest that the holism Woodger sought was of any interest to synthesis actors.[88]

Woodger was born in Norfolk in 1894. A biologist primarily trained in embryology, in 1914 he gained a degree in zoology at the University College London in amniote embryology and physiology. After spending a great part of World War I in the Norfolk battalion in Mesopotamia, he was appointed protozoologist at the Clinical Laboratory in Amarah, where he was occupied in studies concerning plague rats and other local medical issues. In 1919 he went back to England and began research in embryology with his former teacher J. P. Hill. In 1922 he was awarded a University Readership in Biology at the Middlesex Hospital Medical School. There he taught biology, human histology and other medical disciplines; indeed, during his stay at the Middlesex Hospital, he published a textbook for medical students, which, as we will see, anticipated his organismal beliefs.[89] In 1926, with the encouragement of D'Arcy Thompson, he worked in Vienna for a few months with Przibram at the Biologische Versuchsanstalt (Institute for Experimental Biology), also known as the 'Prater Vivarium'.[90] This brief experience must have been important for Woodger's intellectual formation. He mentioned his change of mind in a short sketch of biography that is today conserved at the UCL Archives.[91] Indeed, after 1926 and against his previous exclusive experimental style of investigation,[92] Woodger embarked on a deep philosophical discussion about the fundamentals of biological thinking. When we look at the intellectual context of the Prater Vivarium, we can better understand this shift in his career.

The Prater Vivarium was founded by Hans Przibram and other two friends, L. Von Portheim and W. Figdor, in 1903.[93] After spending a brief period in Leuckart's laboratory in Leipzig, Przibram travelled to Strasbourg to study with the physiologist F. Hofmeister (1850–1922). Once back in Vienna, Przibram purchased the old zoo-aquarium located in the large public park known as Prater. In fact, the old aquarium was sold due to bankruptcy in 1902, and Przibram converted it into an international experimental laboratory.[94] Przibram worked as director of the zoology department, Figdor as director of the botany department, and W. J. Pauli (1869–1955)[95] was appointed director of the physical and chemical department. Finally, P. Kammerer (1880–1926) was in charge of the Prater's terrarium and aquarium. The explicit goal of the institution was, as Przibram himself recalled, 'to tackle all the big questions of biology'.[96] Przibram's vision was to create an institution where many different disciplines and approaches could find common ground – a synthesis between physics, chemistry, biology and physiology. As Pouvreau describes:

> The institution was generally characterized by the opposition to Darwinian and neo-Darwinian theories, accepting, of course, evolution, but denying the idea that natural selection alone could explain evolution and the preformationist conception of heredity proposed by Weismann (1834–1914): the dominant conception was that the organism had to be conceived as a system in an active relationship with its environment, and that the organism's morphogenesis had to be seen as the result of epigenesist processes.[97]

For Przibram, as for many Romantic biologists, developmental processes showed how the whole constantly shaped its composing parts. He was convinced that such a relation of whole/parts could be understood through the help of mathematics: 'the influence of the "whole" on the "part" turns out to be accessible to mathematical processing, grounded on our ideas from exact natural history'.[98]

When Kammerer left the institution following the famous accusations of fraud in 1924,[99] Paul Weiss (1898–1989) succeeded him.[100] Weiss had been a student of Hatschek as well, and his PhD was supervised by Przibram himself. Therefore, his anti-mechanist and holist approach towards developmental problems reflected his background in Vienna. Weiss's studies on regeneration and transplantation, together with discussions on mosaic and epigenesist mechanisms operating in ontogeny, brought him, eventually, towards a *systemic* conception of the organism and its development – a conception that influenced the young philosopher von Bertalanffy, who even in his doctoral dissertation had begun to speculate about how organisms, conceived as entities belonging to a superior hierarchical level of complexity, could be studied and understood in their own individuality.[101] In sum, when Woodger arrived in Vienna in 1925, anti-mechanist, anti-reductionist and holistic approaches to biology, especially characterized from a hierarchical perspective, marked and dominated the Vivarium. To Woodger, the Prater Vivarium offered an exciting environment where experimental biology and philosophical discussions about the nature of life went hand in hand. Weiss's theoretical biology based on system theory and von Bertalanffy's lifelong friendship marked Woodger's intellectual development.[102]

## Organismal Biology: Between Analysis and Relations

> Biologists, in their haste to become physicists have been neglecting their business and trying to treat the organism not as an organism but as an aggregate. And in doing so they may have been good chemists but they have not been good biologists, because they have been abstracting from what is essential to the biological level.[103]
>
> J. H. Woodger

Before going to Vienna, Woodger had not been ignorant of the organismal tradition or of holistic approaches to biology. As a student of Hill at UCL, he may well have apprehended the long embryological tradition characterizing organismal attitudes (from Huxley to the Scottish school). The discussions he had with Ian Dishart Suttie, a Scottish psychiatrist who was one of his comrades during his stay in Mesopotamia, could also have introduced some elements of the holistic tradition.[104] Whatever the case, from 1924 Woodger was a sure supporter and admirer of Haldane's functional biology and organismal physiology. His medical textbook, published in 1924, fully demonstrates that. First, he introduced the book with a quotation from Haldane about the strict relation between function and structure, and how this relation brought further knowledge about what is normal and what pathological in medicine.[105] Second, the whole organization of the textbook reflected the typical organismal approach. It started with the whole organism, its function and organs, then treated the organization of tissues, and finally discussed the properties of the cell; this emphasized to young students that the whole organism's organization comes before its constituents. Third, the conception of the organism and its relation to the environment mirrored Woodger's holist beliefs as taken from Haldane. As Woodger wrote:

> Biological problems always involve a threefold relation. First as we saw at the beginning, the living organism is the seat of ceaseless activity, it is always *functioning*; secondly, its functions are always manifested through some material *form*; and finally, these functions, sooner or later, directly or indirectly, have reference to the surrounding of the organism, that is to say, to its *environment* ... Just as the organism is more than the sum of its material parts, so it is more than a bundle of functions. Just as its parts are organized and unified, so are its functions in reality not separable but all interconnected and coordinated, so as to result, in the living animal, in one great function – the behaviour of the animal as a whole.[106]

When Woodger arrived in Vienna, he was fully receptive to the novelties and new directions that organismal approaches were taking. However, one central notion emerged from his stay at the Prater: the notion of hierarchy. Such a concept required clarification, and Woodger would spend several years in pursuit of this. Organic hierarchy would represent Woodger's original contribution to organismal biology; it would unify several problematic notions: heredity (inborn) or environment (acquired), genetics and embryology, living organization, synthesis and analysis in biology, and a general critique against reductionism and mechanism.

Woodger introduced the notion of hierarchy and its related issues in his early *magnum opus*, *Biological Principles*. However, before setting up the theoretical elements making an organic hierarchy, he proposed his own ideas about biological explanations. Indeed, his epistemology reflected the ontological fact that organisms were hierarchical systems where causes and effects were indissolubly interrelated. Any biological explanation required four conceptual elements: 1) the things or objects requiring an explanation, 2) the ways through which those things can be analysed, 3) the relationships in which those objects stand to each other, and 4) the relationships in which those objects stand to each other and which can be also analysed (reduced) to simpler relations. Now, in biology Woodger conceived two kinds of explanation: explanation by analysis, and explanation by relation.[107] Biological analysis involved, in turn, *perceptual analysis* (specific distinction of characters), *manual analysis* (the anatomical distinctions such as organs, tissues and cells, etc.), *physiological analysis* (the study of the mutual relation of anatomic parts) and *chemical analysis* (the study of chemical compounds as related to life's processes). Although, as he taught his medical students, the different forms of analysis were central in biological or physiological explanation and essential in biological investigations, biology was essentially a science of relations. In biological phenomena, *relations* between organisms and their environment, between organisms and other organisms in the environment, and between an organism and its parts were essential to understanding life's 'mechanisms'. And yet, according to Woodger, the concept of *relation* was strictly related to the notion of living organization itself: 'Biological explanations ... always involve far more than simple analysis but will have to deal with the complex relations between the various *relata* revealed by the analysis. Thus a theory of biological explanation requires a preliminary study of what we are to understand by organization, a question which, curiously enough, is rarely discussed in biological books'.[108]

Woodger would never give a clear-cut definition of 'organization' because he thought living organization was a result of organic relations among lower and upper levels of organic complexity. In other words, organization was all that emerged from dynamic hierarchies. It was an irreducible *fact* of experience. Biologists start their inquiries with such fact. But, Woodger continued, organization implies two kinds of hierarchies, as we can empirically observe – hierarchies concerning what Woodger dubbed *relata* (composing parts), and hierarchies concerning relations among *relata*: 'the organism is analysable into organ-systems, organs, tissues, cells and cell-parts. There is a hierarchy of composing parts or relata in a hierarchy of organizing relations. These relations and relata can only be studied at their own levels and not simply in terms of the lower levels since these levels do not constitute *unit* relata'.[109] But organization, as the manifestation of hierarchical relations, entailed also a temporal and spatial element: 'a living organism is analysable into a hierarchy of parts in a hierarchy of relations, but neither its parts nor their relations are unchanging since it is differentiated temporally as well as spatially'.[110] Organization emerged from the spatial relations among hierarchies in temporal differentiation. As we will see, Woodger's scepticism about genetics and his criticism of gene concepts did not derive from his 'excessive empiricism',[111] but from his pervasive notion of the organism as a hierarchical system and his critical views on the notion of 'causal determination' in embryology and genetics.

## Hierarchies

Between 1930 and 1931 Woodger published three articles in a new journal, the *Quarterly Reviews of Biology*. The journal was founded by Raymond Pearl (1879–1940) in 1926, and it was intended to be an intellectual platform for book reviews and discussions related to historical or philosophical issues in biology. Pearl was an American biologist who had studied biometry with Karl Pearson in England; once back in the United States, he became a widely known popularizer of science, especially on issues concerning population control and eugenics policies. He admired Woodger's philosophical attitude and helped to support his career as main referee. When Woodger applied for a chair in social biology at the London School of Economics in 1930, Pearl wrote a glowing letter of recommendation.[112] Even though Woodger did not get that job, Pearl's admiration for him was further demonstrated when he invited the young scholar to publish a large manuscript in the *Quarterly Reviews*. The manuscript had been divided by Woodger into three parts, and it treated one of the hottest topics of the time: the relationship between embryology and genetics.

The notion that connected both embryological and genetical studies was the concept of the organism as an integrated system of hierarchical levels. In the first part of the manuscript, Woodger began to define and describe in formalist language an organic hierarchy. His treatment, based as it is on an axiomatic system, is sophisticated and complex, but it is essential to an understanding of Woodger's reformulation of the notion of heredity as the expression of developmental tendencies. A hierarchical order was first characterized as a whole integrated entity *W*, analysable into composing members *M* and *m* belonging to two different sets: the set *L* (levels) and the set of assemblages *A* (set of *M* and *m*). Now, both *L* and *A* were correlated by a fundamental relation *Rh* so that every member belonging to a specific level *L* was analysable into an assemblage *A* in which every component in *A* was in *Rh* relation to *A*. The relation *Rh* defined therefore an irreducible level of complexity, although other forms of relation were conceived too: relations *Rl* among levels and relations *Ra* among members *M* and *m* belonging to specific assemblages *A*. For instance, a cell was analysable in cell-part assemblages, and cell-part assemblages could be analysed in term of chemical compounds; yet, although a relation between different levels could be established in principle (so to analyse the cell in terms of chemical compounds), there was not, in Woodger's scheme, a fundamental lowest level *explaining* or *causing* all higher ones: 'We thus have to get out of the habit of regarding only the supposed "ultimate" components as "really real". Otherwise it is quite arbitrary to stop at cells, or genes, or molecules, or even atoms. We are required to do justice to each level in the hierarchy of levels'.[113]

Woodger defined the concept of hierarchy as applied to the notion of organism more and more analytically. Organic hierarchies not only involved different relations and levels, but they also entailed diverse kinds of hierarchies; division, spatial and genetic hierarchies described, through their own specific relational properties, all possible forms of cellular organization. Yet diverse hierarchies were not only composed of things and their various relations, they were also characterized by temporal and spatial parts; to Woodger, a temporal slice or event was always associated with a spatial part. However, he further distinguished between *components* and *constituents* of spatial parts: the former was defined as assemblage *A* of level *L* (i.e. the nucleus is a component of the cell or the cell a component of tissues); the latter was thought as a contingent part without any relation to the level or hierarchy (a beef-steak is a constituent of the cow). To Woodger, an organism was a four-dimensional entity; in fact, the temporal dimension defined the kind of relations that the whole system established in different temporal slices between levels *L* and between members of assemblages *A*. As a consequence, living organization was defined in terms of relational properties; an organism was alive if some specific relational properties between levels and members were established.

Components, organized according to specific relations in a whole *W* divided in levels *Ln*, characterized, among other things, the differences among organisms. What biologists did was to compare different wholes (for instance *W* with *W*1) or analogous slices of related wholes. So, while taxonomists compared wholes of diverse hierarchies, embryologists compared analogous time-slices of related wholes, and geneticists compared characters of an adult organism with related characters of another: 'We might, for example, be comparing a particular shade of gray under the determinable colour of the skin of the adult rabbit A, with the particular shade of brown under the determinable colour of the skin of the adult rabbit B'.[114] Of course, the differences, observed from different perspectives by taxonomists, embryologists and geneticists, reflected both *intrinsic* (or immanent) and *relational* properties that an organism, during its development, exhibited during different temporal slices. Although intrinsic and relational properties could be translated as heredity and environment, Woodger preferred to avoid the word heredity altogether because he considered it vague and misleading; as he had warned in his *Biological Principles*, 'from the very diverse definitions of this word [heredity] given in biological literature it seems clear that the word stands for an indefinable abstraction, and in the interests of precision it could be banished from scientific terminology with advantage. It is a vague term borrowed uncritically from common sense'.[115]

The distinction between intrinsic and relational properties in a hierarchical system allowed Woodger to reformulate the sharp opposition between hereditary or acquired characters, and between inherited or environmental factors. Indeed, intrinsic properties could be thought of as acquired relational properties during development, once ontogeny was defined as the progressive increase of the different relations in an organic hierarchy:

> If two cells, which are assumed to have 'equal' nuclei behave differently in the same environment, we should say that they differed intrinsically in their cytoplasm, since their relations are supposed to be the same. But that intrinsic cytoplasmic difference may have been acquired in consequence of relational differences during development, and would therefore be an acquired relational property. But since it now persists in spite of changed relations (since by hypothesis both cells are in the same environment now) we should have to call it an acquired intrinsic property.[116]

With the aid of the notion of organism as hierarchical system, Woodger hoped to reconfigure the whole debate between embryologists and geneticists and, therefore, between preformationists and epigenesists. Variations between hierarchical systems (so organisms) were due to both intrinsic and relational properties; however, there was no neat division between 'inborn' and 'acquired' properties. Intrinsic properties were, in truth, 'persisting' relational properties as 'acquired' by the hierarchical system as a whole. In the end, Woodger's theoretical framework left open the way for a Lamarckian interpretation.

Yet, as we will see in the next section, his new theoretical approach met the approval of most developmental biologists and was quite critical about genetical sciences. To Woodger, whereas genetics dealt with classifications, comparisons, variations and differences, 'heredity' was the science dealing with immanent and relational properties as manifesting during the different temporal and spatial slices characterizing the whole organism's development. In sum, to him, genetics was not the science of heredity.

## Immanent and Relational Properties

> Thus chromatin takes the place of Descartes' God as the 'controlling mechanic'.[117]
>
> J. H. Woodger

One of the most important legacies of Woodger's stay in Vienna was his acquaintance with animal regeneration phenomena; indeed, initially, he had planned to go to the Prater Vivarium to study transplantation in worms.[118] Przibram was internationally recognized for his experiments on animal regeneration and transplantation, and Woodger certainly benefited from that. Among his papers, conserved at the UCL Archives in London, is a draft of a translation from German of Przibram's 'Regeneration und Transplantation im Tierreich' (Regeneration and Transplantation in the Animal Kingdom),[119] a paper that Woodger probably drafted after his return to England. The phenomena of transplantation and regeneration offered Woodger powerful theoretical support for his organismic approach and his arguments against genetics and any preformistic view. To him, Przibram, Weiss[120] and later Spemann had shown, through their experiments, that hereditary determination depended on intrinsic properties, spatial and temporal relations: 'In such an organism as a newt, in which a high regenerative ability is retained throughout his life, what is regenerated depends not only upon the spatial relations of the regenerating tissues but also on the temporal slice of its history which has been realized, in other words what is regenerated depends on the state of development to which the organism has reached'.[121]

According to Woodger, the experiments on animal transplantation, which involved cutting specific animal or embryos parts and pasting them into the bodies of other animals or embryos, demonstrated the plasticity of the organism's reactions, both in the parts and the wholes. In fact, only rarely were the transplanted parts strictly determined at the beginning because, once they were pasted, their activities changed according to the new 'environment'. The activities of transplanted parts always depended on the activity of the whole system in which these new elements were inserted.[122] Furthermore, both regenerative and transplantation phenomena provided excellent instances of normal development. Then, to Woodger, development itself appeared to be quite incompatible with Mendelian genetics. Indeed, development, given its extreme plasticity and variability (likewise regeneration and the activities related to tissue transplantations), could not be controlled directly by gene activity.[123] In other words, because development was seen by Woodger as an open, regulative and plastic process resulting from intrinsic and extrinsic factors, it could never be ascribed to fixed or determined hereditary particles:

> Development is a process in which with temporal passage new spatial parts come into being all with the same genetic endowment. As development proceeds they do

> not lose 'factors' but selection is made from their possibilities and this depends on their mutual relations, on the relation of the whole to the environment, as well as on their immanent endowment. It is because of the pervasiveness of the earlier modes of development that an organic part from an embryo of one species can be transplanted to a different place in another embryo of a different species. *And it is because of this pervasiveness that they do not admit of analysis by Mendelian crossing, but they are not any the less dependent upon immanent factors ... in the course of development part-events are elaborated whose characterization depends partly on this immanent endowment and partly on their mutual relation to other part-events. There can be little doubt that some of these immanent factors are intimately linked with the events known as chromosomes, and that they have an 'atomic' character. But this atomicity is also in some way overcome by the organization of the whole since it is the characterization of the whole which depends on them and cannot be interpreted as the outcome of the doings of unrelated cells.*[124]

However, like Lillie before him, Woodger did not simply dismiss genetical sciences; he only argued that they could not be of great help in understanding development. Geneticists were not interested in the organic process of 'realization' but only in statistical distributions of the final outcome.[125] The real challenge lying before genetical science was neither to explain or elucidate embryological issues, nor to formulate an all-comprehensive theory of heredity; it was rather to gain knowledge about the law-like distribution of characters among individuals belonging to different generations.

The Whiteheadian process-oriented model of development that Woodger presented was totally unsympathetic to genetic interpretations, and the notion of 'hereditary characters' was considered by him old-fashioned and belonging to a 'pre-scientific stage in the history of genetics'.[126] His new terminology and notions based on intrinsic and relational properties restated, on a new basis, the old debate about nature and nurture. In an organic hierarchical system, asking what was natural and what was 'acquired' was equivalent to asking to what extent the volume of a gas was due to its temperature or its pressure.[127] Even if genes existed, Woodger added, they were 'clearly not the things that geneticists are talking about'.[128] Genes could only vaguely denote all that makes different cells; whether what made this difference was a gene, an entelechy or a mnemonic element did not matter as long as it brought new useful knowledge.

## Against Mechanism, Against Capitalism

Although Russell, as we have seen, was concerned about the social and political consequences of materialist, reductionist and mechanist sciences, he never developed an explicit and wide critique of society and political interventions as being based on a false or mistaken science. Woodger did. As one of the main members of the Theoretical Biology Club in Cambridge during the 1930s, Woodger shared the company of some of the most left-wing, radical scientists active in that period in England.[129] The members participating in the meetings – people such as J. B. S. Haldane, J. D. Bernal and J. Needham – were scientists also fully committed to a political cause: socialism.[130] Woodger himself was interested in the influence of science on society in general. As he wrote in his *Biological Principles*: 'the results of scientific enquiry sometimes have application to other aspects of human life, and some of these applications have, and are likely to continue to have, a profound influence upon it'.[131] However, as a philosopher of science, Woodger was interested not so much in technological applications to society but in how biological ideas could be used to support a socialist ideal. Although, to my knowledge, Woodger never published any pamphlet, book or article with explicit political contents, he did speak about these topics in lectures and speeches. In particular, there is a typescript in the UCL Archive about a paper given at the Middlesex Medical School Socialist Society, apparently in 1945. This document, entitled 'A Biological Approach to Socialism', demonstrates how, even twenty years after Haldane's address at the Institution of Mining Engineers, the analogy of the society as a collective organism persisted.

As Woodger presented, the biologist, as any common citizen, is angered by the poverty and inequality that characterize capitalist societies.[132] Now, he continued, some biological ideas could give support to socialism and, at the same time, condemn capitalism – the conception of the organism, for example, which conveyed the idea of association and cooperation among parts for the sake of the whole. Indeed, apart from properties or activities such as respiration, excretion, nutrition or movement, organisms show other fundamental properties such as aggregability, divisibility and diversifiability. By aggregability, Woodger meant that all living things, from multicellular organisms to societies, tend to form units of higher order. By divisibility, Woodger intended reproduction or, at the social level, migrations. And for diversifiability, he referred to Milne-Edwards's division of labour: organisms are divided into functional specialized regions. A characteristic deriving from these three properties is that when a part of an organism was severed from the whole, both the part and the whole behave differently.[133] Single organisms, as well as whole communities, manifest a mutual dependence on each other. To Woodger, the connection between the organism's properties and society was evident.[134]

However, he made clear that with his organismic conception, he intended to go beyond to the Hegelian State doctrine. The dependence of the individual on the whole society was not as high as the case of cells in an organism; the social organism should always be thought of as constituted from free individuals, i.e. human beings freely cooperating, for the sake of both the whole and themselves. Unfortunately, capitalist societies perverted the beneficial relations of part to whole or individual to individual. Society is seen as an arena for individual competition, where any form of true cooperation is thwarted and disrupted by the few accumulating large amounts of goods against the large mass of population.[135]

When Haldane, D'Arcy Thompson, Russell and Woodger participated in the Second International Congress of the History of Science in London in 1931 – a meeting that, as Werskey records, was characterized by a strong Soviet and socialist propaganda – questions about materialism, mechanism and reductionism in biology were not just abstract issues but conveyed a strong political charge.[136] We will see that in the United States, too, the diffusion of organismal biology was accompanied by a similar political charge. In fact, it was not coincidence that W. E. Ritter, as we have seen, chaired the 1931 session at the London congress. However, in the following two chapters I will show how organismic approaches and conceptions were received in the United States, where the organismic metaphor, more than inspiring personal or individual agendas, influenced entire institutions.

# 5 THE AMERICAN VERSION: CHICAGO AND BEYOND

## Ideas beyond Institutions

In the second and third chapters we have seen that in Europe the neo-Kantian tradition, then dubbed organismic biology, spread through different, idiosyncratic ways in each country and was accepted by a number of investigators with very diverse backgrounds. However, in the United States the diffusion of such a tradition was less complex and more straightforward. Even though important figures such as Louis Agassiz[1] had established an important European outpost of research and education on American soil since 1848 – in organizing, among other things, the important Museum of Comparative Zoology in 1860 at Harvard University, an institution shaped according to his bio-philosophy and pedagogic convictions[2] and which, at that time, provided the training of some of the leading nineteenth-century American naturalists[3] – the great syntheses in biology that are part of my story came later and, as I have already mentioned, from other sources. One of the most important intellectual sources was certainly Leuckart and his school at Leipzig. However, as we will see later on, although Leuckart and his school, as well as other European figures and institutions, inspired and shaped the biological knowledge of many American investigators, biologists in the new continent developed and translated what they had learned abroad. Since the very late nineteenth century, American biology – as K. R. Benson and J. Maienschein argue[4] – had its own intellectual space and therefore its own peculiarities. In fact, the American organicist tradition I will sketch in this chapter and the next, like American biology in general, was not a mere copy of the European biology; it retained its own originality. Nevertheless, it shared styles, methods and key principles with continental investigators.

In this chapter we will see that within the US context, persons, traditions, ideas, institutions, research methods, all interacted in different ways. In particular, when we analyse the development of the American organicist tradition, we find it difficult to uphold, without reservation, the idea – current among some sociologists of science – that institutions and structures of power shape, for example, styles of inquiry, ideas, research traditions and theoretical outcomes.[5] For example, Jonathan Harwood, in his authoritative book *Styles of Scientific Thought* (2003), argues that differences in styles of thought depend on the kind of organization or institution in which the investigator works: 'Which problems are defined as central to a discipline', writes Harwood, 'is determined not by an internal logic but by the structure of power in and around that discipline'.[6] Turning to the particular case Harwood analyses, he points out that stylistic differences between American and German genetics reflected different structures of power and interest.[7] German geneticists were more open-minded about different mechanisms of inheritance – cytoplasmic inheritance, for example – than their American colleagues, and they were also less inclined to focus on transmission genetics, keeping an open eye on developmental issues. This was, Harwood argues, partly because in Germany, unlike the United States, scarce resources and low growth of the higher education system prevented the establishment of new institutions in which new disciplines could be carved from broader fields. On the other hand, Harwood explains, 'Rapid growth of resources within an educational system ... makes institutional innovation easier. Instead of competing with better-established fields for scarce resources, new fields can set up independently of older ones. Geneticists who inhabit chairs or departments designated for genetics, for example, will be freer to define their own research agenda'.[8]

Of course, Harwood identifies other reasons underlying the different research styles and diverse approaches to genetics, such as the organization of German institutions and universities – i.e. their external connections with, for example, industry and agriculture – and also the power of the faculty or a single professor over the development of departments. Following the sociological tradition best exemplified by Karl Mannheim, and the institutional approach of Joseph Ben-David, Harwood reduces scientific styles to specific interests, institutions or 'basic intentions'. In such a scheme, ideas and traditions are in fact mere secondary elements of an underlying social structure which directs and filters the intellectual experiences of the social actors. Such an institutional difference between American and German genetic communities reflected, as Harwood explains, two kinds of methodological approach to science: a pragmatic and a comprehensive viewpoint. Americans, in virtue of their history and institutional organizations, were more prone to solving concrete and affordable issues, whereas Germans were more inclined towards wild theorizations. The experiences of German émigrés in the United States confirmed this: the New World held, as Richard Hofstadter claims, a strong anti-intellectualist component,[9] an aspect totally alien to German culture: 'In such a society', Harwood points out, following Hofstadter, 'knowledge was generally valued not as a route to wisdom, but as the means to utilitarian ends. The man of knowledge was respected, not as a culture-bearer in the German mould, but as someone whose expertise could solve practical problems'.[10]

I linger on Harwood's work because it represents, in my opinion, a very clear and convincing model of sociological explanation, one that sheds light on the reasons why different styles of reasoning succeed and thrive. However, although the Harwood scheme may be true regarding its own context, it would be inappropriate if applied to my case study.[11] Indeed, when we turn to my story, we see that even though American biologists such as Child, Conklin, Harrison, Wilson, Whitman, Ritter, Just, etc. worked for the greater part of their lives within American institutions, they still maintained a continental style and viewpoint in biology; in other words, many first-rank American investigators indulged in broad speculations and theorizing, as did their German and European colleagues. Furthermore, even though original and innovative, their concepts and ideas – in general biology, then in embryology, heredity, morphology, evolutionary studies, their methods and experimental practices – were not in contradiction with European standards, nor did their approaches to science necessarily have a pragmatic end. If we look at American biology from a larger perspective – one not based on the Morgan school or Mendelian tradition in general – we see that Harwood's distinction between comprehensives and pragmatics is much more difficult to accept. If a larger picture of the American context relativizes national styles of thought – in showing that there is not always a direct and smooth connection between institutions, styles and ideas – then we could argue that perhaps scientific traditions, ideas and methodological styles can enter people's minds, notwithstanding strong social and institutional filters. I think that my American case study demonstrates that people and ideas may come before institutions and structures of power. In other words, persons, ideas and traditions can profoundly shape institutions and structures of power rather than be shaped by them.

In the next section I will introduce my first case study: the zoology department at the University of Chicago. I shall focus my research on two figures who had important relations with the institution: F. R. Lillie and his pupil E. E. Just. We will see that the bio-philosophy upheld by these figures – and therefore their holist and organismic approaches to biology – had its roots in Europe, in what I have dubbed the Romantic tradition. Then, in the next chapter, I will introduce my second case study: the Scripps Marine Biological Association in San Diego, and the relation between the founder of this institution, W. E. Ritter, and another Chicago biologist, C. M. Child. At the end of my discussion, I hope to have demonstrated that both these institutions were shaped according to the scientific ideas and philosophical conceptions of their founders and employees, rather than the other way around.

Finally, where I have explained in Chapter 2 the diverse connections tying together American biologists with European traditions, in this chapter I shall focus on the Americans' positive programme, i.e., discussions, ideas, theories and scientific results. Once we have established that US bio-philosophies were rooted in the neo-Kantian, Romantic tradition in Europe, we need to understand in what ways they were original, what kind of positive discourse they proposed, and which scientific agenda they supported.

## The Organismal Style at the Zoology Department of the University of Chicago

The University of Chicago opened in 1892. Thanks to the financial backing of J. D. Rockefeller and the organizational skills of a young classicist, W. R. Harper, the institution started, as Newman records, as a 'fully-fledged' university.[12] In fact, in the same year Charles Otis Whitman was appointed Head Professor of Biology and Professor of Animal Morphology in the brand new department of zoology – a department that, as we will see, would represent a very important place for the development of organismal philosophies in the United States. Whitman was born in Nord Woodstock, Maine, in 1842.[13] He received a classical education in local schools, and in 1868, after taking his BA degree from Bowdoin College, he taught classic and modern languages, as well as mathematics, at Westford Academy. It was through E. S. Morse, a zoologist who taught at Bowdoin College and was a former student of Louis Agassiz, that Whitman shifted his interests towards zoology; indeed, Whitman was among the few students attending the unlucky Agassiz's school of natural history at Penikese Island. After this brief experience, Whitman sailed towards Europe in order to work as a PhD student at Leuckart's school in Leipzig. The German experience made him a prominent embryologist, and after two years spent as a teacher at the Imperial University of Tokyo, he concluded his training abroad by working at the Naples Zoological Station in 1881. During the following years Whitman would change and direct different institutions: from assistant zoologist at Agassiz's Museum in Harvard, to director of the Milwaukee Lake Laboratory; and from professor of zoology at the Clark University to the first director of the new Marine Biological Laboratory in Woods Hole in 1888. When he became head of the zoology department in Chicago in 1892, he was already a star: he was the first editor of one of the most important American journals for biologists, the *Journal of Morphology*, as well as founder, together with one of his Chicago colleagues, W. M. Wheeler, of the *Biological Bulletin*.

The presence of Whitman in Chicago was decisive in fostering an organismal approach to biology, both institutionally and philosophically. As Gregg Mitman observed:

> [Whitman's] concept of organization was one that permeated not only his institutional plans but his biological research as well. Organization did not just emerge

> from a combination of cells or faculty members. Nor was it merely responsible for the maintenance of the organism or institution. Organization instead preceded development; it resided latent within the egg as a potentiality of structure passed on through heredity that awaited actualization. Hence, the highest function of organization was to 'create and direct', not 'to adapt and conform'. Organization was thus a guiding force of development and ensured progress along the way.[14]

Looking at the people Whitman appointed and trained, and the kind of works that were published under his supervision, it is easy to see how he shaped the institutions that he directed according to his received ideas and approach to biology.[15] It was thanks to him that figures such as G. Baur, W. M. Wheeler, C. M. Child and F. R. Lillie were appointed there.[16] Whitman used his power as an administrator to create an intellectual environment dominated by an organismic style to biology,[17] a Romantic style that had many similarities with the European tradition he had absorbed during his stay abroad. With Agassiz, Whitman's department represented the second important European pole of research and education on American soil. After Whitman's death in 1910, a young biologist whom he had trained took his place; that was F. R. Lillie.

## Disseminating Organicism: F. R. Lillie

Frank R. Lillie earned his PhD under Whitman's supervision, and Whitman's influence on Lillie's bio-philosophy is evident from the earlier stages of his career. That Whitman was a constant source of inspiration for the young Lillie is already evident, particularly in the first important papers Lillie published on *Unio*[18] and *Chaetopterus*.[19] Although these were specialist and very focused pieces, the central lesson Whitman had advanced in his own classic papers, such as 'The inadequacy of the Cell-Theory of Development' and 'The Seat of Formative and Regenerative Energy', is easily discernible: the whole organized entity we call the organism (or parts of it as the cells), in all its complex functions, could not be understood by considering its composing elements alone, insofar as we require a 'principle of unity' able to explain how those elements are disposed in such an organized manner. In brief, the whole organism comes before its composing elements, and life resides not in these elements but in their peculiar disposition and dynamic organization.

Lillie was born in Toronto in 1870. The son of wealthy parents, he studied first religion and then science at the University of Toronto.[20] In 1891 he gained his BA degree and began working at the Woods Hole Marine Laboratory, where from 1900 he was assistant director under Whitman. Apart from a brief interlude at the University of Michigan and at Vassar College, Lillie spent most of his career between Chicago and Woods Hole. Although a prolific author, skilled teacher and widely recognized figure by his contemporaries, he was also a keen administrator and organizer.[21] In 1908 he became director at Woods Hole after Whitman's retirement.[22] As Newman explains, Lillie was 'the prime mover back of the development of the Marine Biological Laboratory from a small affair housed in a few frame buildings to the present large and commodious laboratory that is now recognized as the leading marine laboratory in the world'.[23] From 1935 to 1939 he also maintained presidencies of both the National Academy of Science and the National Research Council.

A practical man, as Kenneth Manning describes him,[24] Lillie was not particularly inclined to philosophical speculations. Indeed, careful experiments, detailed observations and very limited working hypotheses characterized his scientific enterprise. An expert in general embryology, he is especially remembered for his research on fertilization.[25] He proposed a controversial theory according to which fertilization is essentially a chemical reaction happening between three elements: sperm, egg and a supposed substance Lillie called 'fertilizin', a substance present on the egg's surface which, once 'activated' by the sperm, triggered the developmental processes.[26] This theory was harshly criticized by people such as Jacques Loeb during the first decades of the twentieth century.

Like all the figures belonging to the organismic tradition I am discussing, Lillie linked his organismal biology with his beliefs about heredity and development. From 1906, when he published the aforementioned paper on the embryonic development of *Chaetopterus*, he argued that it was through embryonic development that heredity manifested itself. In other words, it was through detailed observations on ontogeny – and all the careful experiments aimed to divert normal development – that hereditary phenomena become visible: 'I make no apology for entering into details, because there is no other explanation of heredity than a complete account of development, and one cannot describe even a small part of so complex a thing without many words, unless one knows in advance what is essential and what is not'.[27]

Yet the careful investigations on *Chaetopterus* development also supported other important conclusions, notably that complex hereditary phenomena 'cannot be reduced to the operation of any single factor'.[28] The complex relations between the nucleus and cytoplasm as well as the apparent structural differences among chromosomes left the door open to diverse interpretations. What seemed to be clear, though, was that both the intra-cellular Darwinian theory of pangenesis and the Weismannian theory of determinants were substantially 'unwarranted' insofar as 'The whole economy of nature forbids us to believe that each cell possesses arm, leg brain, liver lung etc., chromosomes of which only one class enters into activity in any given tissue, the remainder lying idle'.[29] Although breeding experiments practised with Mendelian methods could help to establish the relation between chromosome-characters and inherited 'proportions' (at that time Lillie could not use the conceptual distinction Wilhelm Johannsen made in 1909 and 1911 between genotype and phenotype),[30] our knowledge about hereditary material might nevertheless advance but little; in fact, insofar as there are no real resemblances between 'phenotype' and 'genotype' but only a possible correspondence, 'any imaginable degree of knowledge of the unit species-characters would not furnish a particle of information as to the nature of the original germinal characters'.[31] To put it in another way, even though we are able – through the observations of Mendelian ratios obtained through breeding experiments – to find a correspondence between specific characters (such as colour or albinism) and chromosome-characters, we still remain ignorant about the true nature of chromosomal structures.[32]

Lillie's recurring synthesis of Whitman's organismal philosophy with heredity reflected a common refrain among the community of organismal biologists. To them, cells were 'subordinate to the organism, which produces them, and makes them large and small, of a slow or rapid rate of division, causes them to divide, now in this direction, now in that, and in all respects so disposes them that the latent being comes to full expression'.[33] If the organism was the primary source of organization and guided complex processes happening at the lower organic levels, morphological characters could not be linked to discrete material hidden within the nucleus of the cell, because any visible character, from the simpler to the more complex, is the result of the whole organization and not the outcome of direct expression from chromosome-character to the species-characters. In particular, the embryological mosaic theory of development – a theory which assumed the existence of predetermined factors narrowly and firmly related to specific morphological parts – had to be replaced 'by the view that there are certain properties of the whole, constituting a principle of unity of organization, that are part of the original inheritance, and thus continuous though the cycles of the generations, and not arise anew in each'.[34]

## Physiological Heredity

The particulate theories of inheritance – as intended by Weismann's followers and as interpreted by Mendel's advocates – appeared totally unjustified in light of the organismal conceptions of biology as seen by some Chicago investigators (but not only there). Anticipating a topic that would become dear to Ritter, Lille pointed out that 'The organism is primary, not secondary', i.e. the organism is not an epiphenomenon of unspecified germinal particles. The organism 'is an individual, not by virtue of the cooperation of countless lesser individualities, but an individual that produces these lesser individualities on which its full expression depends'.[35] Not surprisingly, many issues we found among European biologists, as discussed in earlier chapters, concerned Lillie too. In effect, the organism not only acted as a whole toward its own parts, but any part was also a result of the interaction between contiguous elements and the external environment.[36]All visible morphological characters were therefore the outcome of these dynamic processes happening at different levels of complexity, plus the environment. Given such a framework, Lille could conclude, once again, that 'characters are not due to "unfolding" of the potencies of determinants but are results of morphogenetic reactions between two or more formative stuffs. The "characters" need no more be preformed in the reagents (formative stuffs) in the case of a morphogenic than in the case of a chemical reaction'.[37]

Yet such a theoretical position was often mingled with epistemic prescriptions: not only did the idea that the nucleus contained all the necessary particles representing adult characters contradict the available evidence, it was totally unconvincing as a possible scientific explanation. Nägeli, De Vries, Weismann and Wilson – all were accused of sustaining such a controversial hypothesis: 'According to these writers ... all the characters that are ever to be impressed by the nucleus on the cytoplasm are represented by original preformations in the nucleus. Such a conclusion appears to me to be practically a negation of the evidences of our senses'.[38] After all, Lillie argued, if all the structural diversity we observe in the adult organism was already present in the chromosomes, it ought to be clearly visible and assessable by 'our senses, by variety of behaviour or reaction'. Not only was such a complex structure not observed, but even its supposition contrasted with the known 'laws of chemical combination'. In short, when a hypothesis was neither supported from evidence nor coherent with other reliable knowledge, it should be rejected: 'It seems to me that all *a priori* consideration should be ruled out of court, unless we are willing to transform biology into a branch of metaphysics dealing with potencies and latencies'.[39] The charge against the particulate theory of inheritance was serious; indeed, Lillie believed that all those advocating for these theories spoke a medieval language,[40] accepted *a priori* assumptions, and overlooked the clear evidence: in a word, they broke the elementary requirements of an empirical science.

Lillie's bio-philosophical agenda became more and more clearly articulated in the following years. In 1909 he published a paper in which his philosophical and scientific ideas were exposed in a popular style.[41] In this paper, Lillie presented the embryo as a more or less stable individual on which the environment acted in different ways. Once again, he stressed the importance of both 'intra-organic' and 'extra-organic' environmental factors during the whole morphogenetical process[42] and ascribed to such a dialectic the difficulty for any preformistic hypothesis in heredity: 'an immense part of we call inheritance is inheritance of environment only, that is, repetition of similar developmental processes under similar conditions'.[43] Furthermore, to Lillie, the difference between a particulate theory of inheritance and a physiological theory of heredity appeared evident once these theories were confronted with some relevant observations. For example, the different colours of mammals did not depend on the presence or absence of specific determinants in the germ cells but on a specific power of the protoplasm to oxidize: 'The development or inheritance of colour ... can certainly not to be due to the presence of black or brown or red or yellow determinants in the germ, assumed for theoretical purposes by some students of heredity, but to a specific power of oxidation of the protoplasm'.[44] Hence, to him, all the problems of heredity, variation and organic evolution had to be tackled within a physiological and developmental framework because, as we have seen, understanding how a morphological character developed was equivalent to explaining inherited and multiple expressions.

The particulate theories of heredity that Lillie was criticizing included also what he called the theory of 'unit-character', a theory supported by De Vries and Bateson, among others. In fact, while Lillie remained sceptical about the supposed relation between germ-character and visible-character (genotype and phenotype), he actually questioned the definition of 'character' *tout court*. He regarded the notion, especially as defined by the early geneticists, as in contradiction with all that the physiology of development had taught. Indeed, Lillie felt it important to specify that, contrary to the received opinion of many zoologists or botanists, 'characters' cannot be seen as anatomical stable units[45] because 'they simply represent the sum of all physiological processes coming to expression in definable areas or ways, and they may thus represent a particular stage of a chemical process'.[46] 'Characters' are like 'shells cast up on the beach by the ebb and flow of the vital tides; they have a more or less adventitious quality',[47] Lillie concluded, quoting Thomas Huxley.

In revealing the depth of Huxley's analogy, Lillie was expressing his organismal faith: if we ignore the nature of vital tides, we also ignore why a shell lies in one position rather than another. But if we could understand more about the tide's movements – therefore, by analogy, more about the physiology of protoplasm – we could also shed light on the hidden mechanisms of heredity, for the simple reason that any character has its own developmental history. Mendelian factors – just as any other kind of hereditary factor – thus cannot be conceived as representative units of definite morphological parts; at most, they 'must be factors in the development of the entire organism'.[48] Genes may trigger developmental processes, but they cannot be the direct 'cause' of any morphological character. With many European biologists, Lillie strongly believed that there was a profound conceptual continuity between Weismannism and Mendelism and that both were in contradiction with organicism.

## Embryos, Feathers, Genes and Hormones: Lillie's Epistemic Pluralism

Lillie, unlike many other figures belonging to the tradition I am describing, did not attempt to formulate a large synthesis of genetics, embryology, physiology and evolution; rather, he looked for a mutual 'ceasefire' between respective disciplines in the name of a methodological pluralism. Indeed, during the first years of the development of genetical sciences, while Morgan and his students were framing a new paradigm of heredity, some embryologists and cytologists in Europe and the United States felt the need to propose a convincing synthesis of studies on inheritance and development.[49] However, this attempt never succeeded, and the clash between two different scientific approaches about the understanding of hereditary mechanisms ended in favour of the geneticists, who, as Scott Gilbert emphasizes, were able to translate the greater part of embryological issues into genetical terms, thus undermining the pluralism advocated by Lillie.[50]

Not surprisingly, Lillie seemed to be quite sensitive to these themes. In 1927 he published a paper in *Science* in which he evaluated the extent of the separation between these scientific disciplines: 'Since Weismann, physiology of development and genetics have pursued separate and independent courses ... there can be no doubt, I think, that the majority of geneticists, and many physiologists certainly, hope for and expected reunion. The spectacle of the biological sciences divided permanently into two camps is evidently for them too serious a one to be regarded with satisfaction'.[51] In the late 1920s genetical sciences were too important to be so easily dismissed, as Lillie did in his early papers. Furthermore, prominent scientists increasingly considered genetic knowledge to be an essential element for a deeper understanding of the physiology of development.[52] It was for precisely these reasons that Lillie tried to assess to what extent – given the assumed importance of genetics for organic development – the concepts put forward by the Mendelian school were compatible with the concepts set out by embryologists.

Lillie began by defining the notion of the germ and its progressive unfolding. First, as Delage had argued several years before, the germ was characterized as a simple entity which manifested itself through the irreducible and dynamic duality between nucleus and cytoplasm. Furthermore, it was a physiologically integrated entity which could encompass different units: from a single cell to a polarized individual with one or more metabolic gradients.[53] At this point, Lillie distinguished between two different consequential processes: on one hand was embryonic segregation, which denoted the establishment of definite 'primordia' at the beginning of embryogenesis, and implied the existence of well-delimited regions with determinate developmental potencies. On the other hand, especially during the later stages of development, such potencies 'emerged' in forming definite tissues (histogenesis) and further 'functional differentiations'.

The concept of embryonic segregation was of paramount importance.[54] Echoing von Baer's law of development, it implied a progressive, dynamic and ordered movement from simpler and undifferentiated 'primordia' to more complex and highly specialized structures – a movement ending with the definitive genetic restrictions that anticipated the consequent histogenic processes. While the first stages of embryogenesis were characterized by a quite dynamic and flexible pattern, this flexibility was progressively lost when development entered the phase Lillie called 'closed terms'. In other words, the diverse developmental branches, exemplified by the open potencies, were now limited to just one possible realization.[55] Once the organism had reached this stage, its morphological parts possessed a relative independence from the whole organism, so that – to use an analogy current at the time – the parts increasingly became independent from the tyranny of the whole.

Once he had provided the theoretical framework in which embryologists were immersed, Lillie tried to evaluate what contribution geneticists could make to such a discussion. Lillie held that genetics could certainly shed light on some biological issues. Firstly, to Lillie, geneticists not only had 'brilliantly demonstrated that genes are concerned in phenotypical realization at different stages of the life history, and it is therefore a reasonable postulate that this is true of all phenotypic realization',[56] but they also showed that the expression of many morphological characters could be *Mendelized*, i.e. it could be statistically demonstrated that a relationship between the development of some phenotypes and genes existed. However, Lillie concluded, regarding embryology, genetics could be more misleading than helpful. Indeed, the reconciliation between embryology and genetics was mainly impeded by the enormous conceptual gap that separated a conception of heredity defined as pure reiteration of an individual organism's life (the physiological conception of heredity) and a conception of heredity seen as the statistical recurrence of specific characters generation over generation. A profound divergence, Lillie asserted, instantiated in different methods of investigation, epistemic approaches and experimental traditions.

Secondly, although genetics could give some contributions in understanding the physical and chemical nature of the germ, it was of no help in working out the process of individuation: indeed, as we have seen, this process was conceived as a complex event strongly mediated by external and internal organic environments. Third and finally, genetics could not help to improve a possible definition of embryonic segregation for the simple reason that geneticists held that each cell possessed the same genetic material. In other words, if all hereditary material was more or less homogenous, genetics could not explain how very different and heterogeneous forms and structures were formed at different stages of development. Hypotheses which supposed some alleged complex models of interactive genes meant putting the cart before the horse; indeed, Lillie argued, these invoked processes supposed what they pretended to explain.[57] Lillie concluded that unless some surprising and unexpected progress were made in the near future, the possibility that genetics could be of any help with the phenomena of embryonic segregation was small: 'I don't know of any sustained attempt to apply the modern theory of the gene to the problem of embryonic segregation. As the matter stands, this is one of the most serious limitations of the theory of the gene considered as a theory of the organism'; if any cell possessed the same genetic material, 'the phenomena of embryonic segregation must, I think, lie beyond the range of genetics'.[58]

Lillie maintained a position of principle: if genetics remained as it was – in accepting the particulate theory of inheritance, for example – and assumed the existence of fixed representative characters linked to specific morphological structures, it could be of very little use to embryologists. The chief methods and approaches of genetics left out completely the phenomena in which embryologists were more interested: it showed the *alpha* and *omega*, namely the relation between the gene and its realization (phenotype), but it overlooked what happened in the middle, all the complex universe of causes and effects that lay between hereditary factors and their visible manifestation. Between genotype and phenotype dwelled a black box which geneticists deemed too complicated to open.[59] It was precisely for this reason that the two fields were still destined to remain separated: they looked at different things with different methods and approaches. They revealed dichotomous aspects of the life history, two irreconcilable perspectives based on two different conceptions of biology – one underlined the firm and unchanging characteristics of the organisms, whereas the other emphasized the constant and always dynamic changing of the living matter: 'The dilemma at which we have arrived appears to be irresolvable at present. It is the apparent duality of the life history as exhibited in the associated phenomena of genetics and ontogeny: on the one hand the genes which remain the same throughout the life history, on the other hand the ontogenetic process which never stands still from germ to old age'.[60]

This duality underlined by Lillie was particularly evident in his studies dedicated to bird feathers in his later works, publications that, as we have seen, were particularly admired by D'Arcy Thompson. The morphogenesis of feathers and the development of particular patterns and pigments demonstrated how a genetical approach could be too 'one-sided'. From 1932 Lillie began to explore the complexity of the morphology of the Brown Leghorn fowl's feathers in a long series of papers.[61] The experiments and observations Lillie undertook aimed to establish, in particular, which factors played a central role in the development of the feather forms, patterns and pigments. In other terms, in these papers he tried to answer a typical embryological question: how feathers, with their sophisticated structure and colours, come to be. The methods of investigation Lillie used consisted, generally, in a nice combination of morphological observation and experimental manipulation. The observation of normal feather development was then used as the base to assess how experimental intervention diverted natural morphogenesis.[62] For example, Lillie noted that injections of thyroxin or female hormone had the power to change feather patterns.[63] Moreover, experiments on transplantation and recombination on the papillae (the sections of tissue at the base of the developing feather) showed that the latter controlled general feather development.[64]

Generally, this accurate experimental manipulation led Lillie to consider the relationship between specific substances such as hormones, the growth rates in different feather regions and patterns, and the morphology of the feather. Quantitative doses of hormones provoked some changes in pattern development, albeit these changes were mediated by the different growth rates in different feather regions. Therefore 'Slowly growing feathers react to a smaller dose of hormone than rapidly growing feathers. As rate of growth is a fixed property of the different feather's tracts ... it is possible to produce birds, by means of suitable administration of female hormone, with female feathers in the slowly growing tracts, and feathers of male character in the more rapidly growing tracts'.[65]

Lillie drew some interesting theoretical consequences from these experimental results. It seemed that female hormones had a specific function, such as unlocking 'the potentiality for female development resident in the constitution of the feather germ. The hormone exteriorizes the morphological and physiological characteristics of the female cell; but both these are the attributes of the cell, determined in their entirety and only dependent on the hormone for expression'.[66] In departing from the same physiological conditions and assuming the existence of the same hereditary material in the cell nucleus (and in the feather's germ, to use Lillie's terminology), diverse morphological structures were formed; all differences related to the variation of growth rates, on which, as we have seen, different hormonal quantities could have different influences. As we will also see in the next chapter, we are not very far from Child's theory of metabolic gradients. In the same way that the action of cyanide or alcohol depended on the metabolic rates possessed by the specific region of the planaria's body, the action of female hormones acted in function of the growth rate of the bird's feather: 'The various patterns formed by injections of hormones are fully explained as to their form by the gradients of growth-rate'.[67]

I linger somewhat on these details because they offer an interesting window on what Lillie really meant when he distinguished between genetics and the physiology of development and advocated their difficult synthesis. In the specific case, whereas geneticists could observe the transmission of specific feather patterns, forms and colours generation over generation – assessing small variations, statistical recurrences and distribution of traits in a population – the physiologist tried to identify the mechanisms hidden behind pattern, form and colour formation. From Lillie's standpoint, patterns, forms and colours were not explainable by positing the existence of genes for any of these elements; instead, they were comprehensible by diverting the 'normal' physiological processes and by 'tracking' the causal chain that, starting from a homogenous hereditary substance, brought it towards a heterogeneous dynamic structure.[68]

Once again, it is possible to conclude that physiology, for Lillie, filled the gap between hereditary materials and their dynamic expression, whereas genetics focused on the statistical transmission of traits, i.e. factors totally abstracted from their ontogeny. Genetics and physiology, Lillie argued, were complementary fields that did not contemplate any reciprocal reduction, at least according the knowledge of the time. Any efforts to formulate eventual reconciliations were 'doomed to disappointment', any synthesis irremediably premature, and any attempt to consider one field more important than the other misleading. Probably the wisest option would be to accept their irreducibility and leave things as they were. After all, even this situation would be better than to translate one field into another, i.e. forcing, simplifying or distorting concepts and meanings of different vocabularies, compelling experimental methods into a unique framework, and circumscribing scientific approaches according to one monolithic paradigm. What Lillie stood for was an epistemic pluralism.

## Embryonic Segregation and its Bearing on Heredity

We have seen that from his earlier publications Lillie thought that genetics could not explain how characters come to be; indeed, at most it could describe regularities happening generation over generation. Genetical sciences could only predict statistically whether a specific character would be expressed in the next generation given the knowledge about the characters expressed in the previous generations. To reiterate, during the late 1920s Lillie did not think that genetics was faulty, he only argued that genetical approaches clarified issues that were strangers to embryologists; problems such as ratios and distributions of characters (phenotypes) in different generations or within a population did not touch developmental issues. In fact, as I have previously mentioned, the central concern of the geneticists was not focused on the causal chain characterizing the progressive unfolding of the morphological traits during ontogeny; what happened between genotype and phenotype was enclosed in a black box, a heuristic and pragmatic gap that nevertheless rendered gene theory open to some paradoxes. One of these paradoxes was clearly expressed by Lillie in 1928 in a letter to J. Huxley: 'if you will excuse a paradox ... gene theory is essentially a theory of phenotypes, i.e., something always static for as soon as it changes it is already another phenotype'.[69]

Again, from a theoretical viewpoint Lillie felt the same kind of uneasiness that many embryologists (including Just, Child and Ritter) expressed about the type of scientific explanations geneticists offered: to say that a particular phenotype has such-and-such characteristics because it is causally linked to an unobservable genotype seemed to be a trivial explanation.[70] It is for this reason that Lillie maintained that gene theory is in its essence a theory of phenotypes; if the genotype was derived from repeated observations of phenotypes, then the former was in reality only a representative label to denote the latter. However, when Lillie published the paper 'Embryonic Segregation and its Role in the Life History' in 1929, he tried to formulate a view through which the geneticists' black box could be partially opened and the relation between hereditary factors and their dynamic expressions would be partly explained. Within such an explicative context, the provisional and partial bridge that would link heredity and its morphological expression was the notion of embryonic segregation, a notion Lillie had already sketched in the aforementioned articles, but that he extended further. Again, this was not a synthesis between embryology and genetics; instead it was an attempt to analyse some aspects of heredity as related to development.

The notion of embryonic segregation was compared by Lillie to the problem German biologists called embryonic determination, an issue which involved the pioneers of the *Entwickelungsmechanik* school.[71] However, although the phenomenon of embryonic determination dealt with some general problems of embryology, Lillie's embryonic segregation denoted a more specific and limited process. In general, Lillie portrayed embryonic segregation as 'the process of origin of the diverse specific potencies that appear in the organism in the course of the life history, which express themselves later in tissues of specific structure and function'.[72] In other words, to Lillie, the general process of embryonic development included the particular processes of embryonic segregation. To use an analogy: just as individual words are the prerequisite for the composition of a romance, so embryonic segregation is the prerequisite for the overall embryonic development. For clarity's sake – because Lillie was introducing a new terminology – he furnished a terminological *legenda*:

| | |
|---|---|
| *Segregate*: | The earliest stage of a part possessing specific potency. |
| *Embryonic segregation*: | The process of origin of segregates. |
| *Primordium*: | The earliest morphologically definable stage of a segregate, organ or region. |
| *Genetic restriction*: | The restriction of potencies of segregates due to embryonic segregation, as compared with those of the parent segregate. |

| | |
|---|---|
| *Phenotypic properties or qualities*: | The properties or characteristics of any stage, or part of a stage, of an organism, whether morphological or physiological, *is esse*, as contrasted with those *in posse*, the existing state either morphologically or physiologically considered. |

The general idea that Lillie was describing was that during embryogenesis, any germinal region possessed some specified potencies (qualities), which were increasingly limited or thwarted as long as ontogeny proceeded towards a critical threshold or phase. At the beginning of development, the fates of different germinal areas were open; in other words, they are almost equipotent. However, once development reaches the threshold Lillie calls the 'critical phase', and the process of segregation gradually achieves its ends, the open potencies of different germinal areas reach a 'close term'. In effect, embryonic segregation was conceived by Lillie as a rather specific event in time. And yet, because embryonic segregation might express a broad range of phenotypical characters, it is different from a phenotypic segregation: different morphological characters might be the result of a unique segregate (see *legenda*).

Furthermore, insofar as development, as von Baer had taught, follows a direction from the simplest to more complex structures, from the homogeneous to the heterogeneous, the development of segregates (i.e. the embryonic regions still possessing open potencies) was always related to what had happened in the previous stages; in other words, the nature of a segregate is determined by the complexity of the preceding segregates: 'Segregates may be designed as complex or pluri-potential so long as they retain the capacity of producing subordinate segregates, and as simple or unipotential when this capacity is lost and only one potency remains'.[73]

Lillie conceived embryonic segregation as a dichotomous process; in the same way that geneticists mentioned dominant and recessive genes, embryologists talked about positive determined segregates and negative determined segregates (remainders).[74] More than one potency could be related to a specific germ because developmental possibilities implied a wide gap between potentiality and actuality, i.e. between actualized lines of development and latent ones. This process of segregation was best illustrated, Lillie pointed out, in the observation of cell lineage. In fact, by following the process of cellular self-differentiation and specialization, and, therefore, progressive restriction of cellular potencies, the observer could easily detect a dichotomous progression. The mechanism Lillie had in mind was comparable to the mechanism Weismann and Roux described in their theory of mosaic development. However, unlike the mosaic theory, which implied that at any embryonic cleavage hereditary material was distributed, split and reduced among the cells, fixing their fate, Lillie's hypothesis entailed a quite flexible process always controlled by the intra-cellular environment. Using the example of the development of triton's eye, Lillie showed that the epidermal primordium involved two different segregates: the complex epidermal segregate and the more specific 'close term' segregate, exemplified by the lens. Because specific segregations depended also on the extra-cellular environment and not on fixed hereditary particles alone, the conception of embryonic segregation Lillie defended was an organismal conception of development, unlike the mosaic theory. It was clearly organismal because, ultimately, it was the organism as a whole which controlled all the formative processes during ontogeny, so that any segregate had to be contextualized within a whole, dynamic developmental system:

> In Weismann's theory of development, self-origination is predicated of all segregates; the idea of self-origination is indeed inherent in all determinate theories, and is equally required in all mosaic theories of development, i.e., those theories in which the individual is regarded essentially as the sum of its parts. But the idea of self-origination is foreign to organismal conceptions, such as now dominate the field of embryology, for the reason that segregates are determinably localized within the organism, which thus controls the exact location in which any given segregate is to appear by virtue of certain of its physiological means ... the foregoing analysis of embryonic segregation is based upon an organismal conception of development.[75]

There was another major difference between the mosaic theory of development and Lillie's hypothesis of embryonic segregation: the latter regarded the relations happening among cells as the result of cytoplasmic interactions. Although Lillie considered the theory of nuclear determination important, he argued that further progress was required to shed light on these phenomena insofar as the available 'nuclear theories' – the classic Weismannian or genic ones – were untenable. Jan Sapp has clearly illustrated that in the first decades of the twentieth century, many important American biologists were opposed to nucleocentric theories of inheritance; heredity, according to these biologists, was not an exclusive phenomenon of nuclear substances but also involved – and in a fundamental way – properties and materials located outside the nucleus. Even more importantly, the cytoplasmatic theories of heredity circulating at the outset of the twentieth century highlighted the flexible and dynamic side of organic development. They were veritable epigenetic theories clashing with supposed Weismannian preformationism: 'The individual elementary units comprising a complex organism, e.g., the cells of blastula, were not necessarily predetermined as different parts but could be primarily all alike in constitution. The differences which arose during development were thought to be determined by the action of environmental factors upon whole groups of cells and upon each member of them'.[76] Lillie's hypothesis of embryonic segregation was the outcome of this theoretical organismic context.

However, even though the theory of embryonic segregation was not intended by Lillie as an alternative to genetics,[77] it could not be overlooked by any broad and convincing theory of heredity. The embryological and physiological studies Lillie undertook throughout his long career – from cell lineage and regeneration to embryology and fertilization, from embryonic segregation to the physiology of bird feathers – convinced him of the unbridgeable gulf separating the conception of heredity as an individual life's reiteration and the notion of heredity as the statistical transmission of visible characters. What Lillie demonstrated, in his popular writings and orthodox works, was that all hereditary phenomena could be understood under two complementary epistemic categories – two independent and irreducible classes of phenomena.[78] The physiology of development, in which the theory of embryonic segregation was subsumed, instead looked at how heredity was manifested and progressively expressed during dynamic and complex processes of individual morphogenesis. Lillie did not confuse heredity with development; he thought that development had a lot to say about heredity from a different perspective than genetics.

## Organismal Biology Applied

The big picture Lillie aimed for led him to divide life history according to three general disciplines or phenomena: morphogenesis, embryonic segregation and genetics.[79] These were all different and irreducible fields with their own methods, languages and questions. As we have seen, Lillie's epistemic pluralism brought him to conceive heredity as a 'multifaceted' phenomenon. Different experimental and theoretical approaches revealed different irreducible aspects of inheritance, from development to transmission, from ontogeny to phylogeny – in a word, from genotype to phenotype, without leaving unanalysed any gap between the two. Genetics was not *the* science of heredity, but just one aspect of it. This wide conception of heredity and development was behind Lillie's failed attempt to create a new institute at the University of Chicago during the late 1920s: the Institute of Genetic Biology. Such an institute, which had to formulate a viable alternative to the orthodox eugenics proposals, was meant to provide a research platform where biology could help to solve social problems. The Rockefeller Foundation, interested in making big investments in areas related to human biology, was intrigued by Lillie's proposal. In fact, as he himself introduced the goals of the institute:

> The future of human society depends on the preservation of individual health and its extension into the field of public health; but it depends no less on social health, that is the biological composition of the population. We are at a turning point in the history of human activity – the age of dispersion and differentiation of races is past. The era of universal contact and amalgamation has come. Moreover the population press on their borders everywhere, and also, unfortunately, the best stock biologically is not everywhere the most rapidly breeding stock. The political and social problems involved are fundamentally problems of genetic biology.[80]

Even though Lillie shared with the eugenicists a fear of the apocalyptic consequences of uncontrolled population increase and disproportionate reproduction by the lowest classes, he was completely against the uncritical and blind application of genetics to eugenics: to him, as to many in Chicago and elsewhere, the environment remained too important a factor in human development. Indeed, as Mitman suggests, 'The physiological, epigenetic orientation of developmental biology at Chicago contrasted sharply with the implicit preformationist and deterministic view of development held by geneticists such as Thomas Hunt Morgan'.[81] To Lillie, the individual was much more than his heredity; it was not through the selection of the fittest that a better race could be created, but through the scientific control of development (including environmental influences). It was embryology and not genetics that represented the only salvation for humankind and civilization. However, such a broad agenda required much financial support, which during the late 1920s – with the world economic crisis – could hardly be afforded.[82] Yet the holistic and anti-mechanist approach that characterized the University of Chicago clashed with the increasingly physicalist line professed by the president of the Rockefeller Foundation, Max Mason, and his protégé, Warren Weaver. They preferred to divert most of the available funding towards a new institution in California: the California Institute of Technology (Caltech). There, as is well known, they developed a new form of biology shaped from methods, views and ideas of the physical sciences, a biophysical science that Weaver dubbed 'molecular biology'.[83]

Although the Institute of Genetic Biology never became reality, it shared the convictions, aims and hopes of another institute that during the same years propounded a similar agenda: the Scripps Marine Association of San Diego, an institution that is the subject of the next chapter. Throughout his life, Lillie had been in contact with the director of the Scripps Association, W. E. Ritter. In the very late 1930s, as a retired man, Lillie still kept in contact with him. They talked about scientific as well as philosophical issue in their correspondence, and in one of their letters, Lillie illustrated the deep philosophical consequences of his experiments on bird feather development:

> There is the same aspect of 'foresight' in the development of the feather that there is in the behaviour of the birds, eliminating of course the aspect of 'consciousness'. Can we have objective criteria of foresight whether accompanied by consciousness or not, and if the answer is negative where is the line to be drawn? What kind of science does this lead us to? It must be something more than a mere return to Aristotle, I would suppose. You probably are acquainted with Spemann's conception of psychological analogies in the processes of development. Ever since I wrote my first paper in 1894–95, where the problem is stated without any consciousness of its implication, this question has been in the background of all my scientific thinking.[84]

The same problem of living organization, an issue that had obsessed Kant himself, and all post-Kantian biologists, was the line that ran through all of Lillie's scientific life. When Lillie published a paper in *Science* in 1938, wondering about the future of zoology, he stated, once again, the organismic principles that had animated his entire career: 'Biology is ... committed to a through-going physico-chemical analysis of organic structure and function, but it is not committed to a reduction of its concepts to physico-chemical levels. Biology is an autonomous science in the sense that its problems concern the level of attainments, both historical and functional, of the living organism'.[85] Lillie had inherited this biological agenda from Whitman, who in turn had obtained it in Europe. The same legacy would be absorbed and transformed by one of Lillie's best students in Chicago, Ernest Everett Just.

## A Romantic Embryologist

> Natur hat weder Kern
> Noch Schale,
> Alless ist sie mit einemmale[86]
>
> J. W. von Goethe

> Just had the qualities of genius; nothing whatever turns him aside from his purpose. I have attempted over and over again to get him to conform to the conditions which his race and the nature of university life in America impose. I think now that this attempt was unwise; certainly it was futile.[87]
>
> F. R. Lillie to R. G. Harrison

Just was Lillie's PhD student at the University of Chicago. While today he is mainly remembered as an icon – as a black hero struggling against the racial prejudices which characterized American society in the first half of the twentieth century; as a brilliant but underestimated scientist who failed to obtain wide recognition in his time because of his colour – his scientific ideas have been almost totally forgotten.[88] However, even though his racial background prevented him from fully realizing his aspirations, and even though he spent most of his career at the Howard Medical School – a black institution he despised – his life was an intense adventure. From very humble origins in Charleston, South Carolina, the city where he was born in 1883,[89] he studied at the Kimball Union Academy and then graduated from Dartmouth College in 1907, obtaining *magna cum laude*. In the same year he was hired by the Howard School in Washington, where as early as 1912 he became full professor and head of the department of zoology. From 1909 Just spent his summers at the Woods Hole Marine Laboratory as assistant to Lillie, who agreed to supervise Just's PhD at the University of Chicago, where Just also met Wheeler and attended Child's courses in physiology.[90] In 1916, although Just gained his PhD in an important institution with a famous supervisor, his opportunities remained rather limited; no institution, apart from those dedicated to black students, was disposed to accept a black scientist, however highly trained; Just had no other choice than to remain at Howard.

In 1929 Just had the opportunity to visit Europe. He went first to Naples, where he studied fertilization in the sea urchin at the Zoological Station, and in 1930 he was invited to the Kaiser-Wilhelm-Institut für Biologie of Berlin. Since then, although tied to Howard for funding, Just visited Europe several times, preferring fascist Italy, Nazi Germany and occupied France to American xenophobic society. Although he dreamed of living in Europe, he was never able to get funding to achieve such a purpose.[91] Finally, in 1940 he was working at the French Marine Laboratory in Roscoff when the Nazis arrested and interned him in a camp. He was soon released thanks to the intervention of his German father-in-law, who had some important connections with the Nazis. Hence he was able to get back in America, where once again he had no other choice than Howard. Just died one year later, in 1941, from pancreatic cancer.

As I have already mentioned, despite the existence of a discrete number of articles,[92] the publication of an excellent biography, and the several important recognitions attributed to Just, we do not know very much about his scientific achievements.[93] Indeed, looking at the existing literature, it seems that his adventurous life overshadows his scientific accomplishments. Therefore we know quite well about his relationships with white women in Europe, particularly with Margaret Boveri. We know about his numerous attempts to get funding from American foundations and universities, we know about some of his connections with eminent scientists of his time: Frank Lillie, J. Loeb, Anton Dohrn or M. Hartmann, among others. We know about his difficult life at the Woods Hole Station, his happy life in Naples, his travels in Europe, etc. However, we do not know very much about the contents of his researches. Of course, we know that he worked with Lillie on fertilization, that he was a very able experimenter widely recognized among the Woods Hole crowd, and that towards the last years of his life, he moved to more philosophical and theoretical issues. We also know that during his final years he published an audacious monograph resuming the researches and ideas on an entire life in universities and laboratories.[94] Finally, we know that he explicitly challenged T. H. Morgan in 1935 on his chromosome theory of heredity, a theory which eventually led Morgan to get a Nobel Prize in 1933; and we know about his enthusiasm apropos of organismal and antireductionist approaches in biology. However, what we really do not know in detail are his scientific alternatives. In other words, what did Just specifically believe about the organism? What about heredity, development and evolution?

In the next sections, I will describe and focus on Just's proposals about what he believed was really relevant in biology; I will also introduce his beliefs and hypotheses about heredity and development, convictions which he developed during the last year of his brief career. With Just, we will see that Kantian and Romantic ideas were alive in the United States even during the late 1930s.

## Mutations, Cytoplasm and Evolution

In 1933 Warren Weaver, a prominent American mathematician who at that time was administrator of the Rockefeller Foundation, sent a letter to Lillie asking for some advice and opinions on the scientific and human skills of E. E. Just. The year before, Just had asked for financial support at the Division of Natural Science of the Rockefeller Foundation, and Weaver was gathering as much information as he could get in order to decide Just's application.[95] Lillie answered in an enthusiastic letter: 'As regards his scientific work, he has been easily one of the most productive investigators at Woods Hole for the last twelve to fifteen years. His studies have been characterized not only by their care and precision, but also by a very considerable degree of scientific imagination'.[96] Just was clearly an excellent investigator and first-rank biologist despite his racial problems and his inability to get a position in an important American institution. However, Lillie did not merely discuss Just's skills and problems, he also mentioned Just's theoretical stance about biology, a position shared by Lillie: 'Biology is a subject in which schools of thought develop sharply defined controversies. Dr. Just is a rather ardent adherent when his convictions are fully formed. Thus I would say that he is against the extreme Mendelian school and also the extreme mechanistic school of thought in biology. I quite fully agree with his views on these subjects, so far as I understand them'.[97] It is not surprising that Lillie agreed with Just's bio-philosophy; after all, it was the philosophy that animated his own technical papers since 1906, and it was the philosophy that – though by then changed and adapted to the new discoveries and knowledge of thirty years of researches – Lillie had inherited from Whitman in Chicago.

Even though Just clearly inherited such a philosophy from Lillie during his training and work in Chicago and Woods Hole, it was only during the early years of the 1930s that he published his most controversial papers. Indeed, between 1932 and 1933 he published, in the *American Naturalist*, two articles tackling different issues: heredity and the nature of mutation, then cytoplasm and evolution, all within an organismal framework. In the first paper, Just conceded that phenomena related to Mendelian inheritance could be accounted for in terms of chromosome combinations and recombinations so that mutational phenomena could be linked to chromosomes themselves.[98] However, Just wondered, even though mutations were related to chromosomes, how did mutations happen first? In other words, scientists had established that some external factors such as heat and radiations were able to induce mutations in *Drosophila*; but, Just asked, how did mutations arise in the chromosomes? We will see that such a question, the methods to solve it, and the answer all betrayed a developmentalist and organismal approach. Indeed, Just was not interested in the kind of mutations he could get through external artificial inductions; he was instead interested in the phenomena of cellular regulation that brought mutation to be expressed. That was not a geneticist's issue but an embryologist's curiosity. Secondly, the methods Just deemed necessary to undertake were embryological; in other words, he described the experimental observations highlighting the importance of cytoplasm in the development of the egg. In fact, Just thought, whether non-nucleated fragments of egg developed anyway; whether changes in salinity or temperature affected cytoplasm without relevant effects on nucleus; whether ultraviolet rays provoked variation of the cytoplasm before affecting the nucleus; and whether external substances, such as oxygen or water, interacted primarily with cytoplasm – all induced Just to conclude that 'Nuclear (and therefore chromosomal) behaviour is secondary to that of the cytoplasm ... Cytoplasm determines nuclear behaviour, the chromosomal behaviour then is an expression of a more fundamental cytoplasmic activity'.[99]

Nuclear behaviour was subject to cytoplasmic reactions, and both experimental cytology and embryology demonstrated that mutations had their primary origins both in the cytoplasm and in the complex modification of cells as unit-systems.[100] Unlike Hermann Müller's opinion, according to which mutations were the result of localized rearrangements of genes, Just thought that mutations were due to chemical reactions involving the cell as whole. Eventually, genetic rearrangements were the effect and not the cause of mutation. In short, chromosomes were not the direct source of mutations because they were controlled by factors coming from outside: 'the normal behaviour of the chromosomes is too rigidly mechanical for them to be responsible as primary agents in heredity. Such a mechanism strongly suggests some deep-seated force of which they are the expression. If we compare them to soldiers going through manoeuvres, we must then assume some source of command. Their orderly behaviour is not automatic but conditioned'.[101] However, the fact that a deeper causal level of reactions was required did not hinder the possibility of discovering a chronological causal chain of respective causes and effects: 'It is highly probable that the first effects of the environment manifest themselves on the cortex. It stands as the medium of exchange between the cell's inner and outer worlds. It is first impressed. The superficial reactions in protoplasm therefore come first. And these reactions certainly must affect the whole cell-system'.[102]

In Just's second paper, the importance of cytoplasm for mutation phenomena, heredity and development was extended even to evolution.[103] Nobody, he stated, could doubt the fact of evolution: too many disciplines converged – taxonomy, comparative anatomy, embryology, physiology and biochemistry – towards the facts that species had changed and evolved. However, even though evolution happened, the mechanisms though which organisms change was still a matter for discussion. Furthermore, another riddle waited to be solved: how life emerged from inorganic matter. It was clear, Just admitted, that the first organic being had to be a kind of complex device separated from environmental fluctuations; still, it had to be 'responsive' to environmental changes. In fact, as he argued, paralleling Haldane and the American biochemist from Harvard, L. J. Henderson (1878–1942),[104] Just noted that 'we should not speak of the fitness of the environment or the fitness of the organism: rather, we should regard organism and environment as one reacting system'.[105] All reactions between a new organic entity and environment had to be mediated by a membrane that Just identified with cortical cytoplasm. Any viable organism able to survive and thrive (probably a simple proto-cell or 'protoplast' as ancestor of both plants and animals) had to be divided in an inner and outer spatial differentiation. The outer part, that is, the cell's cortical cytoplasm, was the most important area because all fundamental functions such as contraction, conduction, respiration and nutrition took place there. No organism, Just concluded, could thrive, reproduce and change without the fundamental contribution of the cell surface.[106] Because of the importance of such an area for all living functions, Just believed that even organic evolution, the progressive modification of the species, was due to specific changes happening at the outer cell level. Organisms change because cortical cytoplasm reacts in different ways to an ever-changing environment: 'Animal evolution advanced rapidly or slowly, to a higher or lower stage, depending upon the degree of ectoplasmic behaviour exhibited as contraction and conduction. Animals to-day differ largely because of differences in these two manifestations of life'.[107] Just did not mention Lamarck or any neo-Lamarckian author, although the mechanisms of variation he conceived shared a neo-Lamarckian approach: organisms change because there is a direct, constant and conflicting relation between living beings (cells) and their environment.

Through these preliminary papers, Just was preparing his larger synthesis, a biological synthesis which brought together development, heredity and evolution within an organicist framework. The hypotheses he advanced on each of these issues were based on a theoretical viewpoint that was clearly organismal, anti-physicalist and anti-mechanist. Indeed, one of his repeated refrains was that neither vitalism nor materialism could be accepted, but that an approach regarding living organization as non-reducible data was required. In his words:

> We have often striven to prove life as wholly mechanistic, starting with the hypothesis that organisms are machines! Thus we overlook the organo-dynamic of protoplasm – its power to organize itself. Living substance is such because it possesses this organization – something more than the sum of all its minutest parts. Our refined and particularistic physico-chemical studies, beautiful though they are, for the most part fail because they do not encompass that residuum left after electrons and atoms and

> molecules and compounds even have been studied as such. It is this residuum, the organization of protoplasm, which is its predominant characteristic and which places biology in a category quite apart from physics and chemistry.[108]

In the following and final years of his life, Just would improve his hypotheses and his synthesis. He would even develop his own original interpretation of genetical sciences and their relation with development, an interpretation strongly focused on cytoplasm's functional activities.

## A Bizarre Form of 'Mendelism'

Between 1936 and 1937 Just published two audacious papers.[109] The two articles addressed one of the hottest biological topics of the time: the possible connections between genetics and developmental biology. They were papers that anticipated the broad synthesis he would propose in his 1939 monograph, but they are interesting in their own right because Just offers an overview of the discussions, arguments and common ideas circulating within the American community at that time. First of all, Just, unlike Lillie, Conklin and Morgan, was profoundly convinced that a synthesis was possible; but such a synthesis required a new perspective in which the apparently unrelated phenomena, i.e. heredity and development, were viewed as two complementary aspects of the same phenomenon. In fact, Just argued:

> all development which we know through experience shows without exception heredity, both at the beginning and at the end as well as at every intervening stage of development. The organism does not exist which has its genesis from an egg and at the same time fails to exhibit characters, Mendelian or otherwise. Nor is it possible for Mendelian characters, especially since they are minute and inconsequential, to display themselves apart from the form of the organism. An organism, either as egg, as adult or as any intervening stage of its differentiation, is heritage, inheritor and visible tangible evidence of heredity; its closest resemblance of its progenitor lies in its grossest and essential parts on which appear the inconsequential minor differences which constitute the Mendelian characters. Since neither process reveals itself independently of the other, we can postulate a cause common to both.[110]

I have quoted this passage extensively because it demonstrates that Just did not confuse heredity and development. He was totally aware that some of the most important biologists of his time considered heredity and development two separate things. However, he disagreed because he regarded the whole matter from a very different viewpoint, a perspective that, as I am arguing, was profoundly organismal.

Before displaying his own proposals, Just noted that biologists of his time were divided in two opposing parties: on one side were the embryologists, that is, biologists upholding Lillie's theory of embryonic segregation (a theory I have described in the previous sections); on the other side there were geneticists, or those Just dubbed '*Drosophila*-culturists', who accepted the gene theory exemplified by Morgan. However, both parties presented weaknesses and misunderstandings; Just dismissed and criticized both the theory of embryonic segregation and gene theory, as least as they were formulated by their supporters. The former was subjected to a wealth of objections, whereas the latter was useless in understanding development because, as Lillie had previously stated, if the cell contained homogenous hereditary material, then the phenomenon of cellular differentiation required other causal factors.

Just followed von Baer's embryology (as he explicitly acknowledged). Differentiation was a process of progressive complexification and could be divided into more or less delimited stages. Any process of differentiation entailed a movement of individual cells that, in turn, were differentiated into nucleus and cytoplasm. Now, Just observed, there was firm evidence confirming the fact that during egg cleavage, and during embryonic development, nuclear substance increased: cell nuclei grew whereas cytoplasm decreased. Just felt constrained to conclude that such a phenomenon was caused by the fact that nucleic substance (chromosomes) removed specific elements from cytoplasm. He argued against the common hypothesis of his time, according to which chromosomes affected cytoplasm in different ways;[111] indeed, he claimed the opposite , i.e. that substances were transferred from the cytoplasm to the nucleus: 'I consider that the progressive differentiation of the egg during cleavage is not the result of the pouring out of the stuffs by the chromosomes into the cytoplasm, nor that of segregation of embryonic materials, but more truly the result of a genetic restriction of potencies by the removal of stuff from the cytoplasm to the nuclei'.[112] The theory required visualization: if the fertilized egg is represented as ABCD (where each letter represents one specific potency), then, during embryogenesis, the first two blastomeres would contain, respectively, AB and CD potencies, then, after a further cleavage, A and B and C and D severed potencies. Now, whereas Lillie's theory of embryonic segregation stated that, at each division, the potencies were segregated, Just's theory prescribed a mechanism of progressive 'restrictions' in which, at each division, the cytoplasm lost potencies in favour of the nucleus and, conversely, the nucleus acquired more and more bounded potencies. In short, when the potencies ABCD are transferred from cytoplasm to nucleus, the egg loses its pluripotency; the blastomeres become 'restricted' and cellular differentiation proceeds.

Just claimed that his theory was both supported by evidence and able to explain many unrelated phenomena. It explained, among other things, phenomena such as polyembryony, merogony, asexual reproduction, post-generation of severed blastomeres, animal and plant regeneration and tumour growth.[113] Moreover, it was a far better theory than Morgan's genetics; indeed, Just produced some of the harshest criticism ever written against gene theory during the year of its triumph. 'The gene theory of heredity', Just argued, 'is an ultra-mechanistic rigidly bound concept',[114] and this 'rigidity', he added, is the reason why it is useless to explain cellular differentiation, which is a plastic and environmentally bound process. To Just, in fact, the real hereditary factors lay in the cytoplasm and were expressed only when the nucleic substance appropriated all the chemical elements hindering their expression. Just's theory, in according a very secondary and negative role to the genes, reverted genetic causation: at most, genes removed all those substances obstructing differentiation and the expression of hereditary potencies. However, such a theory was compatible, Just believed, with the Mendelian phenomena observed by the '*Drosophila* culturists': 'the hereditary factors for pink or red-eye are located in the cytoplasm and pink or red-eye results, depending upon the abstraction of red or pink respectively by the genes'.[115]

Just's alternative theory was an improvement on the classic one because, as he claimed, it could explain how hereditary material could affect the cytoplasm; by absorbing stuff, the genes freed cytoplasm to express its hereditary potentialities during embryogeny: 'I relate the origin of the tangible embryo to tangible cytoplasmic changes. The cytoplasm builds the embryo. Then it builds all of it, including characters called Mendelian'.[116] But even though Just posited the cytoplasm as the first mover in developmental and hereditary phenomena, he, as any organismal biologist, regarded the cell as an integrated unit-system in which nucleus and cytoplasm were in constant interaction: 'The protoplasmic system, nucleus and cytoplasm, is structurally the biological unit and acts as such. Then physiological processes inhere in the interplay of the components of this unit, parts of an integrated whole'.[117]

Another important piece for Just's synthesis was thus displayed: with the physiological interpretation of all hereditary phenomena and with the clear recognition that nucleus and cytoplasm are two necessary elements of a unique interacting system, development and heredity could be regarded, Just thought, as two complementary expressions of the same life-history.

## Cells, Evolution and Philosophy

The monograph Just published in 1939, *The Biology of the Cell Surface*, was the outcome of several years of laboratory work and thinking:[118] from his first experiences as laboratory assistant under Frank Lillie at Woods Hole, to his European experience at the Naples Marine Station, then at the Kaiser-Wilhelm-Institut in Berlin, and finally at the Roscoff Marine Station in France.[119] The book was therefore a highly articulate text in which philosophical speculations in biology were mingled with experiments and technical information. Just's commitment to the organicist principles appears evident, even from the first pages,[120] a commitment in which the influence of Yves Delage is unmistakable: living beings are material things; they are formed from inorganic particles, atoms, and composed of chemical substances. No transcendent entity is required to explain biological phenomena, no mystical forces or entelechies need to be invoked; to Just, organisms were open to direct observation through the accurate and skilled practice of the experimenter. However, although life was coextensive with the laws of physics and chemistry, it was not reducible to them. Any reduction was impossible insofar as living entities exhibited behaviours that inorganic bodies did not; such behaviour was exemplified by living organization acting as a unity transcending its physical components. Not only did living organization transcend its compounds, it also implied a constant process of change: 'life is exquisitely a time-thing, like music. And beyond the plane of life, out of infinite time may have come that harmony of motion which endowed the combination of compounds with life'.[121]

With Kant, Just believed that the mechanical analysis had to be sought as far as possible. Measurements and experiments, in which parts are analysed in their own right, should never be stopped. However, with Goethe, Just thought that the biologist should never forget that analysis alone destroys life and, in so doing, frustrates any attempt to understand functional organization: 'The investigator of the living state can and must use physics and chemistry since the living state is a zone in nature and his method of investigation can parallel that of the physical scientist inasmuch as he finally comes to the unit-organization of life. Below this he cannot go, for life is the harmonious organization of events, the resultant of a communion of structures and reactions'.[122] In addition, organisms had to be observed in their normal state and natural environment before undertaking any artificial experimentation because, as Just argued, in the Cuvierian mould, life was living function before structure. In order to comprehend structural organization, the biologist needs to understand how organisms live and are adapted to their contexts.[123]

Just drew on several sources in order to support his organismal philosophy: apart from Goethe and von Baer, he mentioned Whitman, Sedgwick, Child, Lillie, Woodger, Delage, Geddes and Conklin, all responsible, in different degrees and diverse contexts, for putting forward an organismal conception of biology. This conception, as we have seen, he readily applied to his theory of heredity and development; he used it against Morgan's school and employed it as a theoretical weapon against the 'physico-chemical' school headed by Jacques Loeb in the United States. Yet the great synthesis proposed by Just, a synthesis which tied heredity, development and evolution, was based substantially on the idea that the cell surface was the central zone to examine. In fact, Just believed that all cell functions, from the mechanism of fertilization through to any material interaction with the environment, required cytoplasmic reactions of different kinds – reactions concealing the most important problems in biology.[124] Ectoplasm, in particular, played a chief role so that cell multiplication, cell differentiation, cleavage and development were all narrowly related to the complex behaviour of the cell surface (ectoplasm and cytoplasm).[125] As we have seen in the previous sections, even hereditary potencies were contained in the cytoplasm. The nucleus and the genes had only a negative role of freeing cytoplasmic potencies from 'hindrance-expression' substances. Finally, evolution: differences among species were once again reducible, in principle at least, to the actions and reactions happening at the level of the cell surface. The ectoplasm was the substance most exposed to the environment, insofar as it registered and actively responded to it, so that 'species arose through changes in the structure and behaviour of the ectoplasm. In the differentiation of ectoplasm from ground-substance we thus seek the cause of evolution'.[126] Even though Just's synthesis was original, it attracted no followers and, as such, it was forgotten. Certainly, though, it is a splendid example of twentieth-century organismal philosophy as applied to modern biology.

In the next chapter I shall focus on two other forgotten American syntheses: those of W. E. Ritter and C. M. Child. We will also see that, apart from the role played by institutions such as the University of Chicago and Woods Hole, the Scripps Marine Association in San Diego also played a central role in the tradition I am reconstructing.

# 6 ROMANTIC BIOLOGY FROM CALIFORNIA'S SHORES: W. E. RITTER, C. M. CHILD AND THE SCRIPPS MARINE ASSOCIATION

At the beginning of the twentieth century, several promising young American biologists – people such as E. B. Wilson, T. H. Morgan, E. G. Conklin, R. G. Harrison, F. R. Lillie, E. E. Just and, of course, Ritter and Child – endorsed diverse forms of organicism.[1] In one of the most successful late nineteenth-century American textbooks, E. B. Wilson pointed out that 'as far as the plants are concerned ... it has conclusively been shown by Hofmeister, De Bary and Sachs that *the growth of the mass is the primary factor*; for the characteristic mode of growth is often shown by the growing mass before it splits up into cells, and the form of cell-division adapts itself to that of the mass: "die Pflanze bildet Zellen, nicht die Zelle bildet Pflanzen" (De Bary)'.[2] Paralleling Whitman, Wilson concluded: 'Much of the recent work in normal and experimental embryology, as well as that of regeneration, indicates that the same is true in principle of animal growth'.[3] Many of these young investigators endorsed what their teachers taught them: people such as H. N. Martin and W. K. Brooks at Johns Hopkins,[4] and former Leuckart students such as C. Whitman at the University of Chicago and C. Minot[5] and E. Mark at Harvard. As a result, during the very first decades of the twentieth century, what I have defined as Romantic tradition thrived in many leading American departments of zoology.

We have already seen that part of the reason why such a tradition spread in the United States was that many American biologists spent a significant part of their training in Europe, and especially in Germany. Then, after their stay abroad, they were able to get a job in the US and make careers in important institutions – institutions they helped to shape according to the philosophies and experimental practices acquired in the Old World. Both of the figures I will focus on in this chapter, W. E. Ritter and C. M. Child, represented a vivid instance of this tendency. They were the second generation of biologists who had undertaken part of their formation in Europe, and once back in the US they made outstanding careers: Child at the University of Chicago, Ritter at the University of California, Berkley and, afterwards, in his own Scripps Marine Biological Association in San Diego (which relocated to La Jolla in 1905, was renamed the Scripps Institution for Biological Research in 1912, then renamed again as the Scripps Institution of Oceanography in 1925). Although they were specialists in diverse disciplines, they shared an organismal conception of biology, one they discussed in several letters they exchanged, and also during Child's stays at the Scripps Association. In the next sections I shall assess the importance of such an institution for the diffusion of American organicism. I will then introduce the bio-philosophy of Ritter and Child and describe, with the help of unpublished materials and archival sources, the specific methods and experimental practices they adopted; the phenomena they selected; the questions they raised; and the theories they advanced.[6] Finally, we will see how an organicist agenda inspired and shaped questions related to human biology, heredity, development, evolution and eugenics.

## Europe

'The most wonderful view on earth!', scribbled William Emerson Ritter in his diary as he sat on a hill overlooking the Bay of Naples on a sunny autumnal Sunday in 1894.[7] On 13 October of that year, after a long journey through half of Europe – visiting museums, collections and universities in England, France, Switzerland, Germany and northern Italy – Ritter arrived in Naples to spend a short period of research at the famous Dohrn Institution. On that hill he was accompanied by another American, the physiologist Charles Manning Child (1869–1954). Ritter had met Child in Naples, and they since shared a solid and enduring friendship. With Child, Ritter explored churches, museums, archaeological ruins, restaurants and Naples's dramatic landscapes, and of course spent hours in the lab. Even many years later, when they worked in two different institutions, – Child at the University of Chicago and Ritter at the University of California, Berkeley – they remained in very friendly contact.[8] Ritter and Child, like many scholars before and after them, greatly enjoyed the Dohrn Institutions's hospitality and excellent facilities. Ritter, in particular, remained particularly fond of Dohrn. Many years later, when he became director of a zoological station in southern California, he sent a letter to Anton Dohrn himself, who was by then old and ill. He asked Dohrn how to direct a zoological station:

> how much I would appreciate even a brief statement from you of what, after your rich experience, seems to you should be the central aims, (one or a few), of a marine station, and the worst pitfalls to be avoided! ... In what order should they be taken hold of? What are the best means instrumentally, organisationally, and philosophically, to be employed? On these large questions I have thought much and experienced some. You have though and experienced ten times more than I. How I should like and could profit by the results of your thinking and your experience.[9]

To Ritter, the Naples zoological station remained a model to follow, and not surprisingly his experience in Naples was important for the development of his own training and thought. As C. Groeben and M. T. Ghiselin have stressed, Dohrn's bio-philosophy informed the institutional organization he directed: 'Dohrn treated the organism as a coadapted system of interdependent parts, and the Station as he created it has just such organismal qualities'.[10] Once back in the United States, Ritter would manifest and express, in his publications and numerous speeches, his devotion to organicist and holistic views about biology and philosophy. From the first decade of the twentieth century, he considered himself one of the representatives of what he dubbed 'the organismal conception of life', a conception rooted in an old and venerable tradition – one that, Ritter complained, risked being forgotten by younger generations of zoologists.

Naples was not the only place where he came into contact with this tradition. Ritter acquired his ideas from very diverse sources and places – even before his European journey – and particularly from Harvard, where he had been a pupil of the anatomist Eduard Laurens Mark (1847–1946). Mark, like many of his contemporary zoologists, spent part of his training in Germany, at the international school in Leipzig directed by Rudolf Leuckart (1822–98). As we have seen in Chapter 2, Leuckart was an established Kantian organicist, and Mark, like many other students of Leuckart, spread what he had learned in Leipzig. Ritter himself, as reader and interpreter of Kant's *Critique of Judgment*, felt himself the heir of Leuckart's biology, in particular the organicist biology he defended.[11] As we will see, Ritter enthusiastically embraced the Romantic tradition I have described so far. But he was not satisfied with inspiring young biologists to follow his organicist creed through speeches, articles and books; he wanted his ideas embodied and concretized in an institutional setting. Ever since his time as a young instructor at the University of California, Berkeley, Ritter dreamed of an institution paralleling the European zoological stations. When an occasion for founding such an institution presented itself at the beginning of the twentieth century, he did not reject it.

## The Scripps Foundation and its Institutional Policies

During the last decades of the nineteenth century, California's economy had boomed. As Carey McWilliams reports, the population increased exponentially: 'not only has the net increase of population been exceptionally large, but the rate of growth has been fantastic'.[12] Population growth was fostered by extraordinary improvements in the railroads system and was accompanied by phenomenal growth in commerce, agriculture and the oil industry. Economic and social optimism inspired various utopian views about the shining technological future of southern California. Entrepreneurs, made wealthy by diverse emerging industries and commercial activities, intended to be the protagonists in designing the future of such a new and promising land. Utopian projects were widely discussed. Technocratic plans and bold visions of the future were enthusiastically debated. McWilliams dubbed such a form of political utopianism 'Californian crackpotism',[13] one instance of which was Job Harriman's socialist colony, created in the desert of southern California in the early 1900s.[14] Such utopianism was increasingly associated with discussions about the fundamental role of science and its public diffusion.[15]

In this historical context, the Scripps Marine Association was founded in 1903 by the newspaper magnate E. W. Scripps (1854–1926), his sister E. B. Scripps (1836–1932) and Ritter. From its beginning, the Scripps Association was much more than a mere zoological station. Indeed, E. W. Scripps had a very precise agenda in mind, a project embodying his progressivist, democratic and pragmatic political ideals. As he saw it, scientific research was not done just for its own sake, but for the technological and economic improvement of California and the whole nation. Born in Rushville, Illinois, in 1854, Scripps made a fortune lending money to local newspaper publishers, and throughout his life he founded and financed diverse no-profit organizations, scientific institutions and educational projects. Together with J. Pulitzer and W. R. Hearst, he was one of the richest newspaper magnates in the United States.[16] Documents today conserved at both the Scripps Institution of Oceanography and the Bancroft Library at the University of California, Berkeley show clearly that behind the Scripps Association there were converging ideas that supported a well-defined bio-political agenda, an agenda tying together many different, though related, issues: the role and centrality of scientific research for California's progress; the harsh critique of Weismann's biology, Mendelian genetics and eugenics; the elucidation and diffusion of what Ritter dubbed 'organismal philosophy' (which implied a renewed form of German developmentalism); and finally, but no less important, the more general application of scientific ideas to political progressivism and democratic doctrines.

Before the Scripps Association became a reality, Ritter already had a very respectable CV. Born in Hampden, Wisconsin, in 1856, Ritter began his career as a teacher in public school in 1877. Like many ambitious 'northerners'[17] in the 1880s, he moved to California, where he gained a degree at Berkeley under Joseph Le Conte's supervision. Le Conte was an influential American naturalist and evolutionist; trained in Harvard under the tutelage L. Agassiz, he had proposed an idiosyncratic theory of organic evolution synthetizing Hegelian historicism, Lamarckian evolutionism, Comte's progressivism and L. F. Ward's sociology.[18] Following Le Conte's steps, Ritter also moved to Harvard, where he got an MA fellowship to study with the zoologist and anatomist E. Mark. In 1893, after being awarded a PhD from the same university, he sailed for Europe, spending part of his training in Liverpool with the Scottish zoologist and oceanographer W. A. Herdman,[19] and at the Naples zoological station studying tunicates. He completed his European tour at the University of Berlin in 1895.[20] Once back in the United States, Ritter became a zoology instructor at Berkeley; and in 1899, thanks to his advanced knowledge in marine biology, he was given the opportunity to participate in the Harriman Alaska Expedition, a journey organized by the American magnate Edward H. Harriman.[21] During this trip, Ritter was able to gather and observe unclassified species.[22]

It was in 1892 that Ritter and his students began inspecting the coasts of California in order to find a suitable place for a permanent zoological station.[23] Only in the summer of 1903, however, was the decision made to establish it in a locality close to San Diego. Indeed, since 1891 Ritter had been in contact with Fred Baker, a San Diego local physician who, as an amateur conchologist, used to collect seashells around San Diego Bay. Baker first met Ritter during Ritter's honeymoon in southern California, and he convinced Ritter that the San Diego area would be ideal for a new zoological station.[24] They remained in contact, and in 1903 Ritter decided that a small hotel boathouse in San Diego could be used as a provisional marine laboratory. During their first days in San Diego, Ritter and his fellows received visits from some of the local entrepreneurs who eventually became interested in financing a marine laboratory. As Ritter recalled in an unpublished document written in 1915,[25] he first met E. W. Scripps and Scripps's sister during one of these visits. They were extremely interested in the idea of a zoological institute on the shore of San Diego[26] and quickly offered Ritter the necessary funding. By the end of 1903 the Marine Biological Association of San Diego became a reality.

## Between Science and Politics

As a newspaper publisher, Scripps was not an expert in scientific research; however, as a committed philanthropist and skilled entrepreneur, he imagined a democratic scientific institution composed of original investigators working for the general wealth. As Scripps himself described it:

> I wish to gather together at this institution a number of men of strong minds and force who are eager for research work, eager to penetrate the, as yet, unexplored realms of knowledge. I think of this body of men being organized into separate groups, the groups themselves being composed of different individuals, each group working in a special department, and each individual having some special work to do. In some sort of democratic fashion, I would have these men organize and cooperate, governing themselves and governing each other, but all composing one block or army.[27]

Ritter could not agree more. In fact, as he had found during his brief stay in Naples and his trip to Alaska, scientific research had to be organized according to a very definite division of labour, a division set in a democratic framework. For Ritter, scientific research was similar to a geographical expedition whose experts and trained specialists communicated their observations to one another.[28] Furthermore, the kind of institutional organization Ritter had in mind reflected his deep philosophical convictions about what biology was and how it should be pursued. The philosophical ideas supporting such an enterprise represented one of the most important aspects of the Scripps Association.[29]

First of all, both Ritter and Scripps aimed for the realization of what they dubbed 'humanist biology', i.e. a science fully committed to solving human issues and applied to humanist disciplines. The scientific knowledge acquired in the Scripps laboratories had to inform and clarify economic and political problems as well as social and ethical concerns. Biological research had to give decisive support to the theories and practices developed by professional politicians and sociologists as well as philosophers and educators. Letters and documents on the early years of Scripps's activities exhibit such a concern over and over again. Scripps himself pointed out in 1916:

> I take it that it is impossible to begin the study of psychology, sociology, economics, and even the most vulgar politics, at any other point than that where the physiologist begins. That there can be any adequate conception of the mentality of man without a rather complete, if not as nearly as possible, absolutely complete knowledge of the purely physical man, seems to me impossible. Biology, as a science, might be considered the parent stem of all the social sciences.[30]

The fact that the Scripps Association was thought to be much more than just a marine station is clearly indicated in a letter Ritter sent in 1912 to Meyer Lissner, a Los Angeles politician belonging to the Progressive Party.[31] In the letter, Ritter observed that the political and social regeneration of the United States could be achieved only with a new political party – a party which, furthermore, based its policies on scientific advice. The new Progressive Party, Ritter continued, could improve only if it accounted for 'the biological and psychological basis of individual life and social organization, and so will work out a genuine science, theoretical and practical, of society'.[32] For Ritter, the science practised at Scripps had to go in that direction, i.e. to provide scientific underpinning for the right political decisions.[33] The Scripps Association had to supply progressivist politics with unbiased truths taken from objective investigations into human nature. In a long letter he addressed to Scripps in 1921, Ritter was even more explicit:

> It has become quite clear to me that economic, political, social and ethical teachings are so largely futile or even worse, because their foundations are almost wholly lacking in natural knowledge ... Professional economists, statesmen, sociologists, ethicists and historians know next to nothing about the natural history elements of their respective sciences. They are simply uniformed – untutored – in the foundational phenomena

upon which their systems of knowledge and theory rest ... now there is certainly no way to overcome this fatal defect in the humanistic sciences except to underpin the study of them by actual observation upon, and formal instruction about, lower animals of various grades and very backward races of men as they all work out their problems of securing and using nourishment, reproducing their kind, defending themselves against destructive agencies etc. ... I wish to use the Scripps Research Institution for demonstrating the necessity and the practicability of pre-humanistic courses of instruction somewhat comparable to the pre-medical courses now given in all colleges. Though it will probably be better to call the work 'Biology for humanists'.[34]

Biological knowledge not only supported humanist sciences[35] but also had to underpin a positivist view of the future, where technological knowledge, intellectual progress and well-founded scientific investigations went hand in hand.[36]

In addition, the Scripps Association was thought by Ritter to be a pluralistic institution in which both experiments in the laboratory and observations in the natural environment had to be seen as complementary.[37] Ritter was rather polemical on this point. Many American institutions, he declared, were committed to a very 'monomethodic' approach to biology. The laboratory was considered the only appropriate place where reliable knowledge could be accumulated, and experiments were focused on a few model organisms opportunistically selected for specific theoretical outcomes. Nothing like that would happen at the Scripps Association. Indeed, as Ritter explained in 1922 to the Berkeley anthropologist D. P. Barrows:

Field observation alone unquestionably encourages *illy* supported and more or less sentimentally colored generalizations. Unsupplemented descriptions, whether of organisms as a whole or of parts of organisms, produce results that savor more of the collector and cataloguer than of the whole-hearted student of animate nature. The laboratory too singly confided in has still greater danger because a danger more pervasive and subtle. There can be no question that laboratory biology may have much of the stamp of museum anthropology, of library sociology, of scholastic philosophy, and of cloister theology. We must undoubtedly take many, probably most biological problems, into our laboratories for study. But the idea of learning biology proper in a laboratory or a museum is as preposterous as the idea of learning navigation from a toy ship on a mill pond.[38]

The Ritter policy was readily applied at Scripps. It is especially apparent if we consider the work of F. B Sumner on the influences of the environment on inheritance in the field mouse, *Peromyscus*.[39] Appointed by Ritter in 1913, Sumner was the Scripps specialist in heredity studies and a key figure among the Scripps crowd; Ritter's conception of heredity[40] surely owed a good deal to Sumner's findings.[41]

If we look at one of the Scripps surveys, we see the extent of the inheritance studies: investigations involving both lab experiments and field observations. At Scripps, researches in heredity implied hybridization experiments on animal and plants, followed by animal and plants transplantations, in order to assess hereditary 'tendencies'. Yet colour analysis, the study of environmental factors on variations, and the use of birth records in order to evaluate seasonal sex relations also took place.[42] All these investigations were deemed relevant for social issues; indeed, Ritter encouraged Sumner to undertake studies on racial relations among humans – an investigation related to the immigration issues preoccupying US public opinion in the first decades of the twentieth century.[43] Of course, the same policy was also valid for other disciplines, from the biology of the ocean (which implied descriptions of organisms and their classification, migrations, studies of temperatures, salinities, etc.) to climatology of the area; from the establishment of a natural museum and specialized library to the organization of explorations (which included studies on animal distribution for fishing, the study of the ocean's bottom and even anthropological research on the distribution of human races).

The aims that Scripps and Ritter had announced were rather audacious, but the humanist, educative, political and methodological aspects of the Scripps enterprise comprised only one side of Ritter's vision. From its beginning, the Scripps Association was committed to the diffusion of a very precise philosophy of biology, one that Ritter dubbed the 'organismal conception of life'. Such a philosophy of biology implied the rebuttal of other available alternatives – alternatives that Ritter identified with the biology of Weismann, and its more recent successful instantiation, Mendelian genetics. These doctrines, according to Ritter, underpinned a very dangerous and harmful ideology that, in the first decades of the twentieth century, gained more and more followers in the United States: positive eugenics.[44] As I will show in the next section, Ritter and Scripps regarded their new institution as a bulwark against materialist and reductionist approaches to biology and, at the same time, as the place where a new biology had to be formulated: a Romantic conception of biology.

## A Romantic Institution: Between Progressivism and Organicism

> For the Scripps Institution had had a creed, which its members have repeated with child-like faith, following the words of their father-confessor. One of the articles of this creed had been the importance of studying the relations between the organism and its environment. Another has been a recognition of the one-sidedness of either field or laboratory study, considered by itself, and the consequent need of combining the two for a proper understanding of vital phenomena. Still another has been the necessity of employing rigidly quantitative methods, so far as these may be applicable. Finally, the organism itself has been wholeheartedly recognized as having a real existence, in its own right, and not merely allowed a provisional existence, pending its analysis into chemical, morphological or genetic elements.[45]
>
> F. E. A. Thone and E. W. Baley

Ritter did nothing to hide his positioning of his philosophy of the organism at the base of the Scripps Association. He explicitly connected (both in published documents and in private letters) his theoretical convictions with his Scripps directorship. Behind its institutional policies, there were two connected elements inspiring the Scripps enterprise as a whole: the first was the positive agenda instantiated by Ritter's bio-philosophy; the second was the negative programme exemplified in Ritter's persistent critique against materialism and mechanism in biological sciences – a critique mainly addressed to Weismannism and Mendelism. Ritter did not, however, believe that his organismal philosophy was original; indeed, he considered it as a philosophical tradition rooted in very old discussions about the nature of the living organism. The theoretical elements comprising Ritter's organismal theory – i.e. the prominence of function over structure, the unavoidable teleological aspect of biological entities, the rebuttal of vitalism and materialism, the fact that a whole living organism expresses qualities not directly reducible to its parts – were all ideas belonging to what he saw as a venerable tradition stretching back to Aristotle which German and French zoologists revitalized in the eighteenth and nineteenth centuries. In fact, Ritter admitted:

> Both Miss Scripps and Mr Scripps have become convinced, with me, that there is an idea of wide and deep importance to mankind behind this enterprise. In just what sense they conceive that idea to be mine I am not sure; but I know well that it is mine only in so far as I have picked it out from among a large number of already existent ideas and tracked it home more assiduously than has anyone else. It is my idea only by assimilation and elaboration.[46]

This heterogeneous corpus of old ideas formed a doctrine which entailed both ontological theses on the nature of the organism and epistemic theses about the way in which biologists tackle their investigations. The Scripps Association, Ritter argued, is 'an instrument for working out this idea ... that is why I have made so much from the beginning of organically correlated continuous researches'.[47] According to Ritter, the endorsement of such an idea entailed a very definite commitment to the organization of the institution he directed. In 1914, writing to R. P. Merritt, an administrator of the University of California, Ritter made the direct connection between his bio-philosophy, his inspirations and how all these elements would impact on the Scripps institutional organization:

> I look upon the development of the Institution in all its aspects, physical as well as intellectual, as much as part of the working out of my general theory or philosophy of biology as I do the planning for and carrying out any particular piece of research. The fundamental conception that underlies the whole, as all my biological contemporaries know who have taken any interest in the Institution, is that of the 'organism as a whole', or as I prefer to express it of the organismal integrity.[48]

Some of the leading naturalists behind such a conception were, Ritter added, Cuvier and Goethe, in particular Cuvier, 'who found expression in the statement that every organism presents a fundamental "balance of parts"'.[49] Such a philosophy of biology, which began with Kant and Goethe and continued through the French school, had to shape the Scripps Association and the way it worked. First of all, the institution had to pursue comparative work going well beyond the marine organism; as long as funding permitted, the lines of inquiry had to be extended to the whole of living nature. Secondly, very close cooperation among naturalists working in the institution had to be fostered, cooperation aiming for a unified knowledge requiring 'the obligation which every professional man or specialist in any field of activity ought to feel himself under to subordinate to a certain extent his own ambitions as touching his specialty to the interests and needs of the larger whole of which his particular province is part'.[50] In short, in order to satisfy the requirements demanded by an organismal theory, an institution was set up in which a wide and diversified range of research was done and, at the same time, strong cooperation among different disciplines was achieved.

Although the Scripps Association was committed to the working out and diffusion of organismal philosophy, it was also oriented, as I have mentioned, against any form of materialism and mechanism in biological sciences; for Ritter that meant neo-Darwinian biologies. There is no letter more indicative of and explicit about these aims than that sent to Miss Harriman in 1910. Ritter compared his enterprise to that of Miss Harriman's father, E. H. Harriman, a leading entrepreneur and director of the Union Pacific Railroad.[51] Just as Harriman's father rebuilt the Central Pacific Railroads, Ritter claimed, so the Scripps Association was formulating a new philosophy of biology founded on the ashes of neo-Darwinian-Weismannian bio-philosophy.[52] Such a reconstruction was seen by Ritter as necessary as long as Weismannian biology, together with its derivations, had significant social and political consequences. First, according to Ritter, mechanistic and materialist philosophies undermined social progress.[53] He added, however, that there were even more worrying threats: materialist philosophies had informed and influenced mechanistic theories of heredity, which in turn underpinned social and political doctrines extremely dangerous for liberal and democratic societies. One of them was, as we have seen, eugenics. 'Eugenics', Ritter wrote to Scripps in 1914, 'carried out in accordance with the general biological theories held by the men who are foremost in the eugenics movement in this country at the present time, would establish an aristocracy more heartless and insolent than anything the world has ever seen'.[54]

Very probably Ritter's deep interest in heredity studies derived in part from his profound concerns about eugenic policies. He had already widely discussed the issue with one of his colleagues at Berkeley, C. A. Kofoid (1865–1947). Ritter's reflections on heredity will be presented in detail below, but it is important here to outline briefly Ritter's concerns about eugenics because they were strictly related to the Scripps Association's aims. Ritter lamented that insofar as eugenics was based on a false biology, it could never achieve what the movement aimed at, i.e. the improvement of the race. Instead, even in the best conditions, eugenic policies would have no effects at all: 'I would suggest the Eugenics movement if continuing to rest as largely on defective and fallacious biological theory as it now does, were to carry out fully its desire of preventing the propagation of the unfit and of securing the "mating of the fit", the result would be without appreciable effect in checking the supposed race deterioration'.[55] Still, such a defective biology – on which eugenics depended – could also have negative social consequences because, as Ritter explained to Kofoid in 1913, 'the mode of interpreting organic nature prevalent in our generation and applied more consistently by Weismann than anyone else, is largely responsible for the habit now so common of trying to shunt all responsibility for foolish and evil acts off from the individual upon ancestry, or the race, or society'.[56]

The supposed ideological threat hidden in the Weismannian discourse on heredity was a staple of Ritter's general thinking. In 1919, in his large monograph *The Unity of the Organism*, a book in which he developed his own conception of heredity, he returned to the ideological risks of *Germplasm* speculations: Weismann's theory, Ritter thought, drawing on an idea dear to the American paleontologist H. F. Osborn (1857–1935),[57] is a fatalist doctrine similar to many theistic philosophies of fatalism insofar as the individual is conceived as totally impotent in respect of his/her unchangeable hereditary substance.[58] The Scripps Association, together with the researches undertaken there by Sumner, had to be at the forefront in order to provide the right biological insights that might underpin a sound eugenics – a positive eugenical ideology compatible with democratic and liberal principles.[59] In the end, Ritter regarded the mechanist and reductionist conceptions of heredity[60] – as expressed by some of Weismann's advocates and the Mendelian prophets – not as unsupported scientific hypotheses but as ideological doctrines with pessimistic social outcomes. Ritter discussed his ideas on heredity with many relevant figures of the time – with Whitman, E. B. Wilson, T. H. Morgan and others – but he remained attached to the idea that genetical thinking was so paradoxical, speculative and unscientific that it concealed strong ideological elements.

Given Ritter's background, we should not be surprised at Scripps's institutional endorsement of organismal approaches to biology. Ritter, as we have seen, was a pupil of Mark in Harvard, and he studied and worked in European institutions where he had had close contacts with continental bio-philosophies. His lifelong profound interests in European philosophers and naturalists made him an indirect scholar of Aristotle, Kant, von Baer, Goethe and Cuvier.[61] Furthermore, his constant relations with organismic American figures such as Whitman,

Lillie and Child, his regular contacts with European biologists and philosophers (among others, H. Bergson and L. von Bertalanffy), and his acquaintance with the theoretical biology of E. S. Haldane, D'Arcy Thompson, E. S. Russell and H. J. Woodger[62] had to reinforce his organismal conception of biology. Finally, Ritter's close friendship with Scripps added a significant political and ideological dimension to his institutional enterprise. In Ritter's opinion, European organicist philosophy was a perfect match for American political progressivism: a liberal and democratic creed which strongly shaped the Scripps institutional organization.

In the following section we will see what precisely comprised Ritter's organismal theory, and also how such a doctrine required a conception of heredity in neat contrast to Weismann's germ plasm and Mendelian inheritance.

## Ritter's Organismal Doctrine

> [W]holes are so related to their parts that not only does the existence of the whole depend on the orderly cooperation and interdependence of its parts, but the whole exercises a measure of determinative control over its parts.[63]
>
> W. E. Ritter and E. W. Bailey

Ritter's organismal doctrine required that organisms be considered the fundamental entities any biologist should start with, just as physicists started their investigations with atoms, and chemists with substances. Furthermore, Ritter thought that any scientific inquiry in biology had to begin with a direct observation of organisms in their environment because, as he argued in an article published in *Popular Science Monthly* in 1909, 'The data of biology are living plants and animals. These are what nature presents. To these we must always go in order to a beginning at any investigation.'[64] Ritter prescribed observation before experiment, broad knowledge of the organisms in relation to their environment before any planned investigation was performed in artificial settings or laboratory. Like many organicist figures before and after him, Ritter believed that organism and environment could never be neatly separated except from an abstract perspective.

Secondly, Ritter believed that biology certainly depended on other sciences such as physics and chemistry; yet he held that living things transcended mere physico-chemical organization. After all, he contended, the evidence spoke against a simple equivalence between physico-chemistry and life in organisms: 'The presumption that biological phenomena may be adequately treated in terms of chemistry and physics ... leads inevitably to a forcing of evidence.'[65] Ritter used many scientific examples taken from various disciplines in order to support his philosophical arguments. Atoms in modern chemistry and chemical substances (atomic combinations) have different properties; the former 'are small bodies imagined to constitute visible substances', whereas the latter are the visible result of the dynamic interactions of atoms. Now what Ritter wanted to argue is that our minds are made in such a way that 'every object in the world must be treated *on its own merits*'.[66] This implied that we are forced, given our epistemic limitations and the external constitution of the world, to consider atoms and substances as ontologically different objects (even if related); in short, substances are not mere epiphenomena of atomic interactions, and atoms are not 'more real' than substances. They are two complementary expressions of the same reality which deserve separate investigations and treatment. Thus the physical sciences cannot in any sense 'raise the question of absolute reality', and as a consequence, just as substances are not less real than atoms, so organisms are not less real than their composing parts.[67]

Thirdly, to Ritter organismal doctrine did not entail the acceptance of vitalism. Indeed, as his friend Child similarly put it, presupposing the existence of transcendent powers acting on organic matter was not a better option than pure reductionism. Vitalism was open to objections similar to those advanced against Weismannian reductionism: with all its metaphysical presuppositions, it prevented any possible factual investigation and experimental confirmation.[68] Organismal doctrine appeared as a viable third way between vitalism and mechanistic reductionism. But such a way required the acceptance of a dogma: living organization is an irreducible reality, and any biologist had to deal with this fact. As Ritter argued, 'life ... is the sum total of the phenomena exhibited by myriads of natural objects called living because they present these phenomena. To understand any organism it must be studied as a whole and in all its relations'.[69] In other words, morphological parts *function* for the whole, and those parts are only comprehensible, by the biologist, as parts of a functional whole.

Ritter thought that his doctrinal system was well supported by a wide and interesting historical narrative of biological sciences. In the 1919 monograph, which was a result of several lectures he had given at Berkeley and the Scripps Association during the previous decade, he gave flesh to his philosophical ideas by reconstructing two antagonistic historical traditions, exemplified on the one side by those who thought that 'the organism is explained by the substances or elements of which it is composed', and on the other side by those who believed that 'the substances or elements are explained by the organism'.[70] The former belonged to the 'Lucretian school', whereas the latter fitted into the 'Aristotelian school'.[71] Ritter analysed the Lucretians first. As 'elementalists' (reductionists), they found their nineteenth- and early twentieth-century champions in Bichat, Schwann and Weismann. Schwann, in particular, stressed a molecularistic epistemic approach to all living phenomena, offering a mere *knowledge of organisms* rather than a *theory of the organism*: 'To explain organisms is, according to this theory of knowledge (elementalism), to reduce them to their elements, and it is nothing else'.[72]

Although Ritter deemed the history of elementalism important and the methods upheld by its members significant in retrospect, the organismical standpoint was certainly more fruitful and historically successful. From Aristotelian zoology, passing through biologists like Cuvier, Geoffroy Saint-Hilaire and the French school of comparative anatomy, until the more recent developments of American embryology, the organismal standpoint had never declined. Even though 'Aristotelians' had known a brief crisis at the end of the nineteenth century, owing to the influence and appeal of Schwann's school, a new organismical theory was emerging. The organismal research programme continued through American embryologists and developmental biologists; figures such as C. O. Whitman, F. R. Lillie, E. B. Wilson and C. M. Child fostered a new direction in biology. In particular, the 'physiological' outlook put forward by Child, with his stress on the 'physiological correlations' among parts, introduced a new theory of the organism which highlighted functional, rather than structural, interpretations of living phenomena. No doubt Ritter considered himself, the institution he directed, and Child's biology as defining the forefront of organismal doctrine.

## Heredity as the Study of 'Biotic Genesis'

> Biological elementalism of to-day undoubtedly has its chief stronghold in the realm of heredity. The germ-plasm theory, accepted by probably a majority of biologists as an absolute monarch in the empire of biological thought, was elaborated for the sole purpose of explaining heredity.[73]
>
> W. E. Ritter

Ritter's bio-philosophy begins with organisms, presupposes living organization and requires environment. It demands the study of interactions before substances, processes rather than static models, functions before structures. It is therefore not surprising, following the implications of this discussion, that Ritter considered heredity not as transmission but as organic self-transformation. Ritter was extremely concerned about heredity and its different interpretations. He discussed the topic not only in published articles and books but also in private letters addressed to important scholars of his time. Of course Sumner was one of them, but Whitman, Wilson, Morgan, Pearl and Metcalf were also included in Ritter's constant requests for clarification. As we have seen, the Scripps Association was committed to a crusade against neo-Darwinian tradition, and any hypothesis of heredity supporting such a tradition was scrutinized and rejected. Ritter's rebuttal of Weismannism and Mendelism was an informed one, however. For example, among Ritter's papers at both the Bancroft Library and the Scripps Institution, there are countless notes and comments written on the theories formulated by the most outspoken nineteenth- and twentieth-century figures in heredity and genetics: Galton, Spencer, Haeckel, MacBride, De Vries, Davenport, Crew, Pearl, Pearson, Conklin, Castle, Morgan, Punnett, Klebs, Correns and many others.

In approaching Ritter's thoughts about heredity and development, we can begin with an early document he published in 1900 in the *University Chronicle* issued by the University of California, Berkeley. In the article Ritter suggested the following definition of heredity: 'Heredity is that in the nature of organisms by virtue of which they tend to reproduce their kind, and by virtue of which, also, they are never produced, either as wholes or in part, in any other way than by their kind'.[74] The fact that organisms always reproduce their kind was not so obvious because, Ritter argued, it was a truth only established in the nineteenth century. Indeed, before then beliefs in spontaneous generation dominated the minds of naturalists; from Harvey to Redi and Spallanzani, from Müller to Lamarck, there was not reproduction but generation. Once the idea that organisms always beget their kind was established, the laws of heredity could be properly investigated. Yet studies on heredity showed that any organism is subjected to two opposite universal laws: the first was clearly shown by Darwin in stressing the fact that organisms always differ from one another. Variation represented the dynamic side of the living world. The other law instead characterized the persistent tendency of the organisms to maintain a fixed form, generation after generation; this was seen by Ritter as the static law of the living realm. Such a static property of the biological realm was heredity: 'the force of heredity is a negative force, if indeed it may be called a force at all. What in reality we mean by the force of heredity is the persistence of the type-form that has been engendered by the limitation to the play of forces on the dynamic side of life, through the sum total of conditions under which life exists'.[75]

This presumptive hereditary force was expressed through different kinds of more or less definite characters, and Ritter furnished a rather complex taxonomy for them: individual, specific, generic and class characters. Furthermore, he distinguished between ancestral characters and extra-ancestral characters. The former were characters inherited from the species, the latter were acquired during individual lifetimes. Ritter argued that all specific, generic and class characters were ancestral, whereas individual characters could be extra-ancestral. Now, the question of the inheritance of acquired characters was only applicable to individual extra-ancestral characters. The issue of characters and their possible transmission was further discussed in 1907 with Whitman. Indeed, Ritter explained to the Chicago professor that the specific characters of any organism should be found not only in the egg (as Weismann and his followers asserted) but also in adults as well as at any developmental stage. Whitman and Ritter concluded that hereditary units as assumed by De Vries and Weismann, i.e. as representative characters sequestered in the first developmental stages, were of no help in explaining the expression of hereditary characters, even as a working hypothesis.[76] This debate about 'representative characters'[77] and their presence in the egg was also the subject of a short article Ritter published in 1909 in *Science*.[78] In that paper, Ritter applied his objections to the explanatory philosophy behind Mendelism and its supposed unit-characters:[79]

> The characters of a frog are undoubtedly latent in the frog's tadpole. What is to hinder, therefore expressing or explaining the frog in terms of the tadpole by saying the tadpole carries the characters of the frog? The logic is sound in the statement that the tadpole contains 'frog factors' or 'frogness'. This seems like the method of reasoning that, as somewhere remarked by Prof. William James, would enable Hegel; and his followers to successfully support the hypothesis that men are always naked – under their clothes.[80]

The logic behind 'representative characters' was for Ritter ridiculous, and the Mendelian language was compared to Hegel's doctrine of essence.

One year later Ritter returned to the same issue. He asked E. B. Wilson for clarification, in particular about a well-defined problem: 'When the egg of a white mouse is fertilised by a spermatozoon of a grey mouse, is not the spermatozoon a determinant or determiner of greyness in the resulting hybrid?'[81] Ritter in fact assumed that a sharp distinction between 'determinant' and 'determiner' was required; the former, understood in Weismannian terms, only *represented* the greyness potential quality contained in the germ cell, whereas the latter indicated a causal-chain relation between a specific region of sperm and the development of grey hairs.[82] First, Ritter claimed, we are not sure that chromatin alone is sufficient for carrying all morphological potentialities expressed in the adult stage. Other factors could play an important role.[83]

Second, the fact that there is a representative relation between chromatin and adult characters does not add anything to our knowledge. Indeed, in Ritter's terms: 'hundreds, yes probably thousands of other things than the spermatozoon between the sperm and the grey hair of the adult are determiners of greyness. The hair papillae of the ectoderm, the ectoderm itself, the blood coming to the ectoderm etc, etc, are surely determiners more close at hand, so to speak, than is the germ cell'.[84] Here Ritter introduces his *principle of standardization*, which was one of the pivotal tenets of his organismal philosophy as applied to heredity: any part of the organism during development is a 'determiner' of something, and in the chain of possible visible effects more causes must be invoked: causes lying at different levels of complexity. In other words, to say that x (chromatin or part of it) represents y (greyness) is only a figurative language that should never be reified[85] because our experience teaches that it is an incorrect oversimplification: 'if I examine and describe ever so carefully the spermatozoon of a grey mouse I do not find any grey colour – any "greyness". It is only by a long developmental process in which a great many stages occur describable in terms among which neither "greyness" nor "spermness" can be used except with the hardest kind of forcing'.[86]

Ritter did not confuse heredity and development, however; he simply rejected the theoretical presuppositions hidden behind such a distinction. When, in 1911, he discussed the matter with another eminent Columbia scientist, T. H. Morgan, Ritter had in mind precisely such a distinction. From 1908 Morgan had been engaged in a successful research programme in heredity with some of his pupils, mainly based on the study of chromosomal behaviour of the fruit fly *Drosophila melanogaster* – the programme was to lay the foundations of modern genetics.[87] Ritter knew well the embryological work of Morgan before his enthusiastic commitment to Mendelian genetics, and reproached him for the fact that his own previous embryological work was in contradiction with his genetical results. In fact, the way that experimental embryologists worked and what 'Weismann-Mendelian' advocates thought were seen by Ritter not as two different though legitimate disciplinary interests but as alternative hypotheses.[88]

Drawing on E. G. Conklin and W. Johannsen,[89] Ritter argued that the orthodox conception of heredity took its meaning from other contexts than biology: 'Another weak spot in much thinking about heredity ... is due to the fiction of "transmitting" characters from parents to offspring. This appears to have come from the original meaning of the terms *inheritance* and *heredity*, which have to do with heirship to property'.[90] For Ritter, economic analogy dominated genetics. Yet it was a bad analogy simply because, whereas hereditary characters are the outcome of a complex series of organic transformations, properties and goods inherited dealt with something fixed by law.[91] The analogy did not work because, as he had clearly mentioned to Kofoid, any supposed visible 'character' was the result of an unbreakable developmental chain – each ontogenetic stage, with all its unique characteristics, is caused by its preceding state, so that:

> we can no more describe, or express, or explain the adult stage of a given individual in terms of the egg or germ cell of that individual than we can describe or express or explain one individual man, for example, in terms of another individual man ... the upshot of all this for the logic of the student of biological genesis is that he has to deploy or expand the familiar, unbreakable cycle hen-egg, egg-hen, hen-egg etc. and recognize that a stage chosen at random from any part of the cycle or concatenation is at the same time explained by all stages before it and explanatory of all stages in front of it[92]

A similar argument was restated in letters Ritter sent to many of his interlocutors. All these discussions, though, never supposed or showed any confusion between heredity and development; they rather highlighted the impossibility of talking about heredity without development.[93] With Child and many other organicists, Ritter believed that there were no things transmitted, but only indefinite potentialities expressed at different ontogenetic stages. As he explained in 1916 to Maynard Metcalf, a zoologist at the Johns Hopkins University,[94] determinants or genes could never be the only cause of morphological characters. Genes do not determine anything if not in cooperation with many other causal elements.[95] The only way we could still continue to talk about 'determiners' is to conceive of them as 'starters' or 'initiators' of ontogenetic transformations: 'on this view', Ritter concluded, 'all other factors that enter into the production of the result, no matter where situated, in the cytoplasm or anywhere else, have as much right to be called determiners as the particular factors that initiated the series of changes'.[96] If genes were seen as mere 'starters' or 'initiators' of developmental processes, the final outcome – the phenotype – resulted from a long chain of causes. Hair pigment, for example, could not be 'caused' by that pigment's gene because a pigment's gene, say red, would have been only a starter of a long causal chain that, during development, produced red pigment. What was really transmitted were not discrete things but, as Ritter's friend Child clearly argued, reaction systems.[97]

Ritter had his own alternative. He argued that a scholar of hereditary phenomena should investigate 'the real study of biotic genesis'. This meant that a proper study of hereditary mechanisms implied not only the direct investigation of the germ-cell structure and substance, but also of how the formation of hereditary parts is achieved; it needed observation, comparatively, of how organs or parts of them develop from preceding stages. In other words, any biologist interested in heredity had to follow what Ritter dubbed a 'descriptive ontogenesis'. It is worthwhile to quote the example he gives:

> the lens of the vertebrate eye originates from the patch of ectoderm exterior to the optic globe. The optic globe itself arises by an outpocketing of the primitive brain. Since both lens and globe resemble the corresponding parts of the eye of ancestors near and remote, their development comes under the principle of heredity; and the ectodermal patch giving rise to the lens, and the part of the primitive brain giving rise to the optic globe are mechanisms of heredity; and the whole observable series of embryonic parts which culminate in the completed eye are the only direct evidence for the mechanism of heredity for the eye. So is it with all biotic ontogenesis whatever.[98]

More generally, Ritter defended the importance of development for understanding heredity and evolution. He was essentially a developmentalist and typologist;[99] he never accepted the idea that population genetics and natural selection could explain speciation. Only specific developmental disruptions could really explain evolutionary diversifications. Variation was not due to gene mutations but to alternative trajectories in development.

## Ritter's Synthesis

Nine years after the great biological synthesis that Ritter had formulated in his 1919 two-volume monograph, he published a long article with Edna Watson Bailey[100] entitled 'The Organismal Conception, its Place in Science and its Bearing on Philosophy' (1928). In the article they proposed a large research agenda aimed at unifying several disciplines, both scientific and humanist. Cytology, physiology, genetical biology, respiratory biology, neurology, endocrinology, psychology and physics (in their terminology, sciences of inanimate nature) were all to be unified under an organismal conception of biology – a conception that had much to say to philosophy and epistemology. This paper is particularly interesting for at least two reasons. Firstly, the organismal research program appears as a collective international commitment. The philosophy of Whitehead[101] towered behind all individual scientific fields, and each discipline had its own organismal representative figure: people such as Whitman, Lillie and Child were, of course, at the forefront, but other more or lesser known figures were also enlisted, including E. Rohde, L. W. Sharp, J. C. Smuts, C. A. Kafoid, J. S. Haldane, C. S. Sherrington, G. H. Parker, G. N. Lewis, C. J. Herrick, E. S. Russell and C. L. Morgan. Secondly, the paper introduced a synthesis between heredity, development and evolution in an original way.

Ritter in fact endorsed Morgan's *Emergent Evolution*; he believed that 'Emergent evolution and the organismal conception applied to living nature are the same thing looked at from different directions. "Emergent evolution" is what that "same thing" is called when the origin and development of living beings are the central interest, while the "organismal conception" is what it is called when their morphology and physiological functioning are considered'.[102] However, in common with many biologists at that time, Ritter's views about evolution were quite anti-Darwinian. As he had explained to Scripps more than fifteen years before, organic evolution was far more complex than expected, and natural selection alone could never explain the evolutionary modifications to which organisms were subjected.[103] Still, in an unpublished paper titled 'Biology Greater than Evolution', Ritter analysed the issue in a broader context: 'Evolution ... is the central, the really great idea of the modern world, and hence is the mainspring of modern motive and action. Yet we have seen that the doctrine of evolution is proving unsatisfactory'.[104] Even though no one could doubt the fact of evolution, the causes behind such a phenomenon are partially unknown. Arguably, Ritter claimed, many causes should be invoked – natural selection, hybridization, mutation, isolation, orthogenesis and environmental forces[105] – although the acceptance of evolution was in no way dependent on a full knowledge of its causes. Furthermore, any evolutionary account had to include individual development in all its complexity. Ritter complained that the dominating philosophies of evolution pretended to explain morphological differences only through a modification of the whole race or species, overlooking individual processes of growth. He compared such a mistake to the genetical theory of heredity, which pretended to explain adult organisms supposing the existence of definite particles within germ cells.[106]

We have seen that according to Ritter, no real separation between heredity and development could be proposed, and in 1928 he updated his ideas on this. In fact, he now accepted as a hypothesis the existence of genes, but he considered them entities formed by organisms during ontogeny, in the same way as any other organic product such as tissues or organs.[107] In other words, genes neither explained nor controlled development; they were effects of it. Then, as Ritter and Bailey explain, 'On the basis of our conception of organic activity as implying reaction systems, to which chemical transformation as well as surface interchanges are fundamental, we can make reasonable guesses about the nature and origin of genes'.[108] Firstly, there were not individual genes but systems of genes, and these consisted in the 'smallest masses into which a portion of the substance of a living organism can be resolved, and still preserve (in the latent state) attributes of the organism'.[109] Secondly, gene systems were produced during the first developmental stages, when the individual was undergoing its first chemical differentiation. During these earlier phases, germ cells were affected from parental 'impressment', conceived as a chemical process that in turn produced a system of genes. In short, gene systems were nothing other than a chemical form of the species 'impressing' on potential offspring. Once these early processes were achieved, embryological development followed, as Lillie had shown with his theory of embryonic segregation. According to Ritter and Bailey, such a hypothesis was not only in agreement with a wealth of embryological evidence, but it also avoided treating genes as 'entelechies' or 'psychods'.

While both ontogeny and phylogeny characterized evolution, heredity and development were unified under a single hypothesis which regarded gene systems as effects of development and not as causes of it. Insofar as development preceded heredity, the only way to study the mechanisms of inheritance was through embryological observations. Embryology remained the fundamental science behind evolutionary studies insofar as the knowledge of individual development gave central pointers in understanding phylogenetic modifications.

## C. M. Child's Romantic Biology

Ritter's private diaries for the period spent in Naples are dense with information confirming his close friendship with Child. Their friendship also involved a profound agreement on bio-philosophical issues: the organismal theory they developed was very similar, though from two different perspectives. Whereas Ritter was much more involved in a deep philosophical discussion[110] on the nature of the organism and the different epistemic of biological investigations, Child based his theoretical speculations on his physiological experiments, which often focused on animal regeneration and physiological regulation. The intellectual connections between Child and Ritter are also explicit in Ritter's publications; indeed, as Ritter wrote in his 1919 monograph, 'Child formulates views of the nature of organisms that agree very well with the organismal standpoint upheld in this volume'.[111] In short, they shared a very similar conception of biology, namely as a science generally studying the irreducible interactions among parts without losing the integrity of the whole.

In addition, as the letters exchanged with Ritter, the Scripps institutional reports and H. L. Hyman's biographical memoir demonstrate, Child preferred spending his summer researches in San Diego rather than in Woods Hole. Apart from the research facilities that Ritter enthusiastically provided (and the rich fauna present in La Jolla), Child had his own community of friends and colleagues who met regularly every summer that he was there.[112] Yet, as I will show later on, Child's ideas and theories were constantly discussed with Ritter and the Scripps staff. The organismal philosophy Child endorsed and developed was not only a result of his training in Germany and the strong influence of Whitman or Wheeler on Chicago's young crowd, but it was also the outcome of Ritter's friendship and Child's frequent visits to Scripps.

On the other hand, Ritter considered Child the expert on animal regeneration and physiological regulation, and he often asked for clarification not only on these topics but also about the theoretical consequences such phenomena entailed. Child, for his part, discussed both experimental and philosophical issues with Ritter, for instance the relation between nucleus and cytoplasm, individuality in organism and organic correlations, the conceptual differences between abnormal and normal growth, development, and many other problems. Furthermore, the Child–Ritter correspondence offers an interesting window on the development of Child's organismal philosophy and gradient theory. One of Child's pupils, Libbie Hyman,[113] recalled that from 1910 Child 'began to perceive that the unity of the organism is a matter of correlation'. Plastic organisms like *Planaria* showed, during their processes of regeneration, an organic tension and correlation among parts: a complex set of partial activities aimed to re-establish the whole reshaping and reconstructing of the parts. 'His search for the mechanisms of correlation', suggested Hyman, led Child 'to the gradient theory which emerged about 1911 and with which his name will always be associated'.[114] What Hyman does not say is that Child regarded his conception of organic correlation as related to other investigations undertaken in Germany some decades before. As Child explained to Ritter in 1909: 'It would not be correct for you to credit me with originating the idea that division of organic systems may occur in consequence of weakening of correlation or its elimination. That idea has been suggested repeatedly by the botanists: Goebel, especially has expressed himself to this effect in some papers. Among the zoologists Roux has repeatedly asserted that as the organism is weakened its parts become more independent'.[115] Indeed, Child considered his hypothesis of physiological correlation as an extension of these pioneering investigations undertaken in Germany during the second half of the nineteenth century.[116]

However, as I have mentioned before, Child's organismal philosophy, unlike Ritter's, was closely related to his experiments and results on animal regeneration and developmental regulation; in other words, his organicism ran in parallel with the formulation of his theory of metabolic gradients. As we will see later on, Child's recognition of the importance of organic physiological correlations during ontogeny fostered and reinforced a new and articulate definition of the organism, which in turn supported his synthesis of heredity, development and evolution. Child, like Ritter, never underestimated heredity. As Child clearly stated in 1911, if we define heredity as 'the capacity of a physiologically or physically isolated part for regulation', then 'the only way in which we can discover anything concerning this capacity is by allowing the *regulation* to occur under the most various and carefully controlled conditions, i.e., we can investigate and analyze the problem of heredity only with the aid of development'.[117]

Child, with Ritter, Lillie and then Just, did not confuse heredity and development but considered them inseparable, because he painstakingly denied any particulate theory of heredity: any theory of heredity made sense only within an organismal framework insofar as the organism was considered the real unit of inheritance. Child believed that every new reproduction was indeed a new epigenesis, a new formation never reducible to any supposed germinal substance carried from organisms generation over generation: 'wherever a new whole, a new "individual" arises in the organic world, there we have before us the problem of heredity'.[118] In fact, Child thought that what was transmitted or inherited were not factors, elements, ids, genes or anything we can define as discrete molecules, but 'potencialities' that, for any specific generation, were 'actualized' in dynamic relations between internal and external environment: 'Heredity is the sum total of the inherent capacities or "potencies" with which a reproductive element of any kind, natural or artificial, sexual or asexual, giving rise to a whole or to a part, enters upon the developmental process'.[119]

Child first formulated, in a broad and accessible way, his organismal conception in two books published in 1915: *Senescence and Rejuvenescence* and *Individuality in Organism*. Both works were, as he explained, a record of his researches and results during the previous fifteen years, investigations which aimed to understand the most simple reproductive processes that implied 'the whole problem of the organic individual, its origin, development, physiological character and limiting factors'.[120] Although the books tackled similar problems from slightly different perspectives, the rhetorical order Child imposed on his arguments is the same: a general definition of what is an organism, and investigation and assessment of all the experimental and theoretical implications such a definition implied – a strategy that he frequently used in his articles, as well.

In the following pages, it is not my intention to provide a comprehensive description of Child's biological thought overall. What I will instead show is how Child connected his scientific conception of the organism (including both epistemic and ontological issues) and his large synthesis of development, heredity and evolution. It is interesting to note that Child's scientific programme was not based on a single model organism, i.e. *Planaria*.[121] Instead, his theory of metabolic gradients was based on a very broad comparative approach – a method that included several organisms with diverse characteristics. For instance, in contrast to Morgan and his school, Child believed that a comprehensive theory of heredity had to be linked to various organic forms. Experiments and observations had to be performed on plants and animals of different complexity.

## Life, Organisms and Colloids

Like Ritter, Child was a strenuous enemy of Weismann's germ plasm and the Mendelian theory of heredity. He used large parts of his books and articles to criticize both theories, considering them a direct result of vitalist traditions and mystical speculations. Once he had established the total inadequacy of these doctrines, Child tried to build up a consistent physico-chemical organismal theory able to meet the requirements of a proper scientific investigation. However, before undertaking the hard job of defining what an organism is, Child was concerned with the definition of what life is. Indeed, for him, if life did not consist in any process based on or dependent on particular vital molecules, it was 'the result of many processes occurring under conditions of a certain kind and influencing each other'.[122] Child conceived life as a dynamic process involving different levels of organization. He regarded organisms not as material things but as physico-chemical systems in constant change. The organism was not an individual self-contained static unity; it was a dynamic unity that 'itself determines, constructs, and harmonizes ... it channels and develops a characteristic morphological structure and mutual dynamic activity in mutual relation to each other'.[123] Organisms work and function according to different modes of physiological 'correlations' which involve mechanical contiguities, transportation of substances through different parts and conduction of chemical signals.

For Child, however, as a proper physico-chemical system, the organism was composed in a quite particular substratum that many chemists of the first decades of the twentieth century identified with colloidal systems.[124] Drawing on Heinrich Bechhold's *Colloids in Biology and Medicine* (1912), Child concluded that organisms were made of colloids. As a provisional definition of the organism, he proposed that 'A living organism is a specific complex of dynamic changes occurring in a specific colloid substratum which is itself a product of such changes and which influences their course and character and is altered by them'.[125] If a colloidal substratum was cause and effect of the organism's changes, this implied that organic structure was strictly related to organic function, so that 'function produces structure and structure modifies and determines the character of function'.[126] Against W. Roux – who distinguished between structural formation and functional development – Child argued that during development, functions were not superimposed on a developing structure; rather, the organism built itself from its own functional activity. Unlike a man-made machine, which does not work until its structure is accomplished, a living mechanism 'functions' from its first formation. In echoing Goethe, Child concluded: 'organism is always functioning while it is alive: life is function ... in no case does the organism begin to function only after its construction is completed ... it constructs itself by functioning, and the character of its functional activity changes as its structural development progresses'.[127] Development for Child was a functional process characterized by a highly complex set of ordered mechanisms, implying growth (increasing amount of organic substance) and reduction (constant elimination of substances accumulated previously). Both processes had to be related to the activity of metabolism.

But growth usually implied another related process: differentiation. This process was, for Child, characterized by a 'structural complication ... different regions of the cell, different cells or cell groups, become different from each other and from the original undifferentiated or so-called embryonic condition'.[128] Differentiation followed a given plan that directed the countless cells in undergoing diversification and specialization.[129] In general, just as growth was often associated with cell differentiation, organic reduction could be associated with dedifferentiation (even if not necessarily). The phenomenon of dedifferentiation was totally rejected by Weismann and his followers because Weismann's theory of development was conceived as a one-direction process, i.e. only differentiation occurred. However, Child added, dedifferentiation was a very common phenomenon occurring in lower organisms with great capacities for regeneration, and resulted from 'the breakdown and elimination of the differentiated substratum or certain components of it, and the synthesis of new undifferentiated substances from nutritive material, as well as by the reversal of the reactions which occurred in the differentiation'.[130] Regeneration showed that development could be a reversible process involving some regressions as much as progressions. In short, Child considered the processes of growth and reduction – both characterized as differentiation and dedifferentiation – associated with a dynamic change of metabolic activities. The fact that metabolic rates were involved in all developmental processes opened the door to a broad scientific investigation, a very large comparative study on many lower and higher organisms.

## Poisoning, Cutting, Comparing; Framing an Organismal Method of Investigation

Child explored two possible ways to prove the existence of metabolic rates involved in the processes of growth and reduction: the direct and the indirect method of susceptibility. Child observed that every organism had a certain degree of susceptibility to specific chemical substances, and this degree was related to the organism's metabolic conditions. In fact, Child aimed to show

how different parts of the organism responded differently to varying quantities of chemical substances (such as ethyl alcohol, chloroform, cyanides and other kind of narcotics). If a narcotic was administered with the intention of killing the organism (the direct method), differences in susceptibility along the body axis were established. In particular, equal concentrations of lethal substances administrated to *Planaria* of different ages showed that there was a quantitative relation between the time *Planaria* required to disintegrate and their age: given an equal concentration of narcotic, older *Planaria* lived longer than younger ones. When the narcotic was administered with the intention to harm organisms (the indirect method), an inverse quantitative relation was observed: younger specimens lived longer than older ones. Child interpreted these data by supposing that younger specimens, with their specific metabolic rates, had a higher susceptibility when exposed to lethal narcotics and, for the same reason, were less susceptible when exposed to non-lethal quantities: 'the animals which have the higher rate of metabolism and die earlier in the concentrations of the direct method live longer than those with the lower rate in the low concentrations used for the acclimation method'.[131] In other words, the experiments seemed to demonstrate that the metabolic rate is higher in young specimens.[132]

Child performed other experiments in order to assess the relation between metabolic rates and susceptibility. In cutting small pieces from *Planaria*, he was able to show that regenerative processes were related to higher rates of metabolism's activities because, as in the previous cases, *Planaria*'s pieces exhibited the same behaviour as the whole specimens: 'The pieces with the higher rate of reaction disintegrate earlier than those with lower rate'.[133] In other words, pieces manifested higher susceptibility to the direct method, but again lower susceptibility to the indirect method. The existence of these differential susceptibility rates were interpreted by Child as proof that the whole organism, as well as its isolated pieces, had to be seen as an individual dynamic entity composed of 'metabolic gradients'. In a letter sent to Ritter in 1915, Child wrote that 'metabolic gradients' were essentially:

> a means of getting things going in an orderly manner. It does not necessarily persist throughout life and in any case undergoes modification and complication. Every region of local growth, for example, gives rise to a gradient which may be temporary or permanent. In the adult body of mammal or man there must be millions of these gradients, ranging from those in individual cells, e.g., in epithelian cells, to whatever traces remain of the primary body gradients. In the simpler organisms the primary gradients undergo less modification than in the higher forms and in many cases the gradients persist from the beginning of development throughout life ... undoubtedly gradients determine relations of dominance and subordination in function as well as in development.[134]

Metabolic gradients present along the axes of these specimens were, as Child claimed, of a quantitative nature. The experiments he performed in order to change the developmental paths in *Planaria* and sea urchin embryos – experiments aiming to inhibit or delay gradients – demonstrated the existence of specific quantitative rates along the main axes. In particular, in exposing designated regions of frog embryos to narcotics, Child showed that it was possible to retard or inhibit 'developmental processes in the posterior region of the body while in the anterior region development proceeds more or less normally'.[135] Child believed that the production of monsters or abnormal organisms through manipulation of gradients demonstrated that metabolic rates could be studied in a quantitative way insofar as they developed in a quantitative way. Child based his findings on a highly comparative method involving experimental analyses of different forms.[136] In fact, he always questioned himself: did other organisms behave similarly? Are there relations between invertebrates and vertebrates? What differences? What similarities?

To Child, what seemed to relate all the organic forms was the fact that every organism develops according to an ordered principle.[137] This order, which represented one of the most important characteristics of life, characterized what Child dubbed 'organic individuation'. In echoing von Baer's embryological teachings, Child believed that individuation involved two different morphological types: the radiate type, in which the process of individuation begins from a central region; and the axiate type, in which individuation starts from different areas. In general, the process of individuation follows an axial polarized gradient which was inherited during sexual or asexual reproduction. The existence of these axial gradients was easily observable in many different organisms: ciliate infusorians, *Hydra* and other species of hydroids, *Tubularia*, sea urchins, starfish, fish, salamanders and frogs. Furthermore, the existence of these polarized axes along the body of all these forms indicated that there was a relation between metabolic gradients and axes, i.e. a definite polarity arising from a metabolic gradient along a definite axis. In fact, according to Child's conception of the organism, 'the axial gradients are the basis of polarity and symmetry'.[138] Embryology, in particular, showed that relationship clearly. The gradients are visible from the first stages of development. Child suggested that during morphogenesis, the origin of organization and regulation could be observed during early axial formation because each stage of embryonic development followed these polarized axes.

Even though comparative observations and experiments performed on the *Planaria*, tubularia, corymorpha, strawberry, fish embryo and chick embryo proved the existence of these axes, they also demonstrated the existence of dominant and subordinate regions as related to metabolic gradients. As we have seen, the possibility of inhibiting or delaying the development of specific parts through controlled experimentation – therefore inhibiting or triggering the emergence of new axes and new gradients – suggested to Child that organisms were shaped from definite tensions: 'the region of highest rate becomes the chief factor in determining the rate of other regions, and since the rate thus determined is higher in regions nearer to it and lower in those farther away, a gradient in rate results'.[139] Child's hypothesis of metabolic gradients and tensions between dominant and subordinate areas explained different phenomena that were otherwise unintelligible – for example, the fact that in *Tubularia* the apical region was independent from other body parts, whereas the behaviour of basal ends depended on what happened to the superior regions. In *Planaria* such a phenomenon was even more evident: 'An isolated piece of the planarian body is not capable of producing at its anterior end any parts characteristic of levels anterior to that from which it came, unless a head forms or begins to form first', Child noted, continuing: 'any piece ... is capable of giving rise to parts posterior to the level from which it came. In short, anterior regions are dominant over posterior regions in regulatory morphogenesis'.[140] Although the general development of the roots and rhizoids in plants also showed similar patterns, that is 'They are ... subordinate to the individual as a whole',[141] they depended primarily on the dominant upper regions where the gradient had been first established.

Child used all his evidence and subsequent interpretations to display his general hypothesis about development and regeneration. During development, the region with the highest metabolic rate establishes a gradient axis, which in turn guides the development of all subordinate parts of the organism's body. Animal and plant regeneration follow the same path: cutting a small piece from the body of a *Hydra* or *Planaria* established a new metabolic gradient – therefore a new dominant region – which would progressively reduce its metabolic rate as the whole body was reconstituted. As a consequence, reproduction could be interpreted, both in lower and higher forms, as re-establishment of a new metabolic gradient driving the development of a new individual: 'the degree of individuation is dependent upon the rate of metabolism. At any given time of development the higher the rate of metabolism, the higher the degree of individuation'.[142]

In *Individuality in Organism*, Child offered a more schematic representation of his metabolic gradients theory. Imagining a 'spherical mass of living protoplasm' whenever an external stimulus triggered a physico-chemical reaction in a specific area, a metabolic reaction was established. As in the case of spreading waves in a pond, such a reaction spread along the spherical surface while a metabolic gradient was taking form. The establishment of these gradients followed, as we have seen, from the formation of a polarised axis which represented 'the starting point of the "mysterious" organization'.[143] Yet the physico-chemical transmission of these 'waves' established regions where the metabolic rate was higher and regions where it was lower, leading to regions of dominance and subordination. Once an axis was established along a metabolic gradient, it could fully or partially persist through different generations so that such an order was transmitted during the processes of reproduction.[144] All correlations among parts, translated from an organismal viewpoint as tensions between dominant and subordinate regions, were considered by Child as 'the foundation of unity and order in the organic individual', and therefore 'the starting point of physiological individuation'.[145]

## Heredity, Asexual Reproduction and Developmental Evolution

The phenomena related to animal regeneration certainly were the most inspiring to Child. As T. H. Morgan had shown with his comprehensive studies of these phenomena,[146] regeneration offered a great deal of material that could be used to understand development. However, for Child, regeneration also opened a large window on reproduction. As he claimed in 1911:

> In the reconstitution of the whole from a part all the essential features of the reproduction of an organism of specific character from a reproductive element of a certain constitution are as truly present as in the development of the egg, though in different form because of the difference in conditions. Whether we call the process regeneration, restitution or reconstitution, whether it involves extensive redifferentiation in the isolated part or is chiefly limited to localized outgrowth and differentiation of new tissue, it is development, morphogenesis, just as certainly as is the formation of an organism from the egg[147]

Comparative observations on the reproduction of lower forms, for example hydroids and plants, demonstrated that asexual reproduction through fragmentation was rather frequent in nature. Such an extended kind of reproduction was seen by Child as a process entailing fragmentation following physiological isolation. Child's notion of reproduction paralleled that of von Baer. In his *Ueber Entwicklungsgeschichte der Thiere*, von Baer had argued that reproduction was essentially a process of progressive independence of a part from the mother's organism: 'the action of reproduction consists in elevating a part into a whole; that, in the course of development, its independence in relation to that which is around it increases, as well as the determinateness of its form'.[148] With von Baer, Child also admitted that 'The history of the development of the individual is the history of its increasing individuality in all respects'.[149] In Child's more sophisticated version, though, the physiological isolation triggered cellular dedifferentiation, which in turn caused the emergence of a new metabolic gradient and, finally, a new trend of cellular differentiation and individuation. In the same way that regeneration followed from the emergence of a new axis and establishment of a new metabolic gradient in severed pieces of *Tubularia*, *Planaria* or *Hydra*, equally, asexual reproduction consisted of cellular dedifferentiation (implying rejuvenescence) and establishment of a new metabolic gradient.

Although Child suggested that there were similarities in sexual reproduction, he also recognized that organisms reproducing sexually reached a higher stage of individuality which prevented any reproduction by budding. In other words, high degrees of cellular specialization and differentiation prevented any process of dedifferentiation and rejuvenation. However, even though the existence of specialized germ cells permitted sexual reproduction, these cells were neither simple nor sequestered at the very beginning of development, as Weismann had suggested. In fact, as Child specified: 'The gametes arise in the course of development like other specialized parts, and like these also possess a definite history of differentiation'.[150] Weismann's hypothesis about the early sequestration of germ cells lacked evidence, Child contended. The supposed separation between somatic and germ cells needed revising. Looking at the whole of the plant and animal kingdoms, invertebrates and vertebrates, Child could find only a few controversial cases fitting the germ plasm theory (the parasitic worm *Ascaris megalochephala* observed by Boveri, the fly *Miastor*, the gnat *Chironomus* and the worm *Sagitta*). Child's attack on the germ plasm theory is withering, and it is worth quoting in its entirety:

> in spite of the complete absence of any trace of early segregation of germ cells in many organisms, in spite of the fact that the egg cytoplasm, not the nucleus, is apparently responsible in most if not in all cases of early segregation, in spite of our ignorance in many cases whether the so-called primitive germ cells really give rise only to gametes, and, finally, in spite of the remarkable conception of the organic world to which the germ-plasm theory leads us – in spite of all these difficulties – the view that these processes of early specialization in the egg constitute a spatial morphological segregation of the independent germ-plasm from the body or soma still finds supporters[151]

After an extended morphological description of different kinds of gametes, Child concluded that 'the gametes are physiologically integral parts of the organism, that they are, like other parts of the organism, more or less highly differentiated cells, and that, like other parts, they undergo differentiation because of the conditions to which they are subjected in the organism and not because of peculiar, inherent properties'.[152] Germ cells were highly complex and did not contain an undifferentiated substance that, during development, underwent differentiation. Child observed that germ cells, both from a morphological and a physiological viewpoint, were among the most differentiated cells of the entire organism.[153]

Child's denial of Weismann's germ plasm theory had important theoretical implications. If germ plasm was nothing other than 'any protoplasm capable under the proper conditions of undergoing differentiation and reconstitution of a new individual of the species',[154] then heredity had to be found elsewhere in the organism. Indeed, heredity, as a broad notion comprising 'the inherent capacities or "potencies" with which a reproductive element of any kind, natural or artificial, asexual or gametic, giving rise to a whole or part enters upon the developmental process',[155] could not be associated with and reduced to some particular substance carried by organisms generation to generation: the whole pantheon of different posited particles – factors, characters, germs, gemmules, pangenes, etc. – had nothing to do with what really happened with the phenomena of inheritance. Heredity dealt with potentialities and activities triggering ordered processes aimed at reconstituting new individuals from isolated parts – whether a piece of *Planaria*, a part of a plant or a gamete.[156] If organisms emerged from dynamic systems of reactions, in which differentiation and dedifferentiation, growth and reduction caused defined tensions between dominant and subordinate regions, then the unit of inheritance had to be the organism itself:[157] 'the original specific reaction system in which the gradient arises is the fundamental reaction system of the species, the basis of inheritance and development'.[158]

As we have seen with Ritter, for Child, if a science of heredity was to deal with individual development, then experiments and observations had to be addressed to the mechanisms causing ontogeny. Animal regeneration, as an appropriate instance paralleling normal development, offered the best materials for solving the riddles of heredity because it permitted infinite experimental manipulations, treatments far more advanced than breeding strains: 'I believe there is much to be learned from these simpler forms of reproduction that the breeding and crossing of sexual forms can never teach us'.[159] In fact, as experiments performed on *Planaria* and other regenerating forms demonstrated, hereditary potencies of different body parts could be 'awakened' by changing their position with other parts.[160] Through cut-and-paste, poisoning and diverting normal developmental processes, hereditary potencies emerged in all their possibilities.

However, if Weismann's hypothesis of a neat separation between soma and germ plasm was rejected, evolutionary change could be explained by invoking neo-Lamarckian mechanisms of inheritance. As Child reported: 'biologists have been slow to admit the possibility of such inheritance (acquired characters), largely because it conflicts with the Weismann theory', but if 'gametes are integral parts of the organism, there is no theoretical difficulty in the way of such inheritance'.[161] With Ritter, Child believed that many causes behind evolutionary change had to be considered or assumed. In any case, if Child's physiological theory of heredity was accepted, evolutionary change could not be caused by a dialectic between structural modification of the germ plasm and natural selection, but as physico-chemical changes of reaction systems during individual development. Organic evolution was therefore characterized by a 'change from a less stable to a more stable condition in the dynamic reaction system which constitutes the organism'.[162] At the same time, genetic characters assumed by geneticists could be equally explained as result of physiological correlations between parts of a dominating whole.[163]

Child hypothesis of metabolic gradients, through which system reactions emerged, unified heredity, development and evolution under one unique framework. Yet Child believed that his physiological theory of heredity was a far better alternative than genetics:

> The apparent independent variation of the characters, the Mendelian phenomena, the association or coupling of characters, sex limited inheritance and in fact all the known phenomena of inheritance can be far more readily accounted for on the basis of different dynamic equilibria. If the organism is a dynamic system, changes in its constitution or in the conditions of the environment may alter its equilibrium and such changes may become evident, now in this character now in that or in a group of characters, according to the nature of the organism and the condition concerned[164]

## Sociological Biology and Developmental Eugenics

Fausto-Sterling and Mitman have nicely highlighted some of the social implications of Child's physiological theory of heredity. Indeed, along with many of his colleagues at the University of Chicago, Child was a progressivist who endorsed social reform policies. With Ritter, he was convinced that the social environment played a central role in shaping human development. Inspired by his own experiments and observations on animal regeneration, he believed that all organisms (including humans) were much more flexible than eugenicists would admit. As he pointed out in 1927: 'We have the best reasons for believing that, within the limits of the hereditary potentialities of the individual, environment and its educational effects are potent factors in determining human character and personality. The fact that it appears at present to be difficult to alter the hereditary potentialities through the action of environment does not justify us in ignoring, or minimizing, the importance of environment for the individual'.[165] Ritter could not have agreed more.

Indeed, as we have seen, the Scripps Association was committed to providing the 'right' biological knowledge for solving social and political issues – in particular a sound biology for eugenics, a biology focused on human development. Year after year, Ritter formulated viable education projects for the 'Foundation of Human Biology'. At Scripps, students had to follow lectures and laboratory courses on 'Stages in the Cycle of Individual Life of All Higher Organisms', therefore a thorough study of all developmental phases happening in Man and his near related species, from germ cell to senescence. Furthermore, probably sometime between the 1920s and 1930s, Ritter even proposed a project for experimental research on child development, involving the selection of twelve children who would be supervised from six months to three years. Special teachers, nurses, dieticians, paediatricians and psychiatrists would be enlisted to carry out didactic experiments. The problems that such a project would address involved both physical and mental child improvement, or in Ritter's words: 'The development of physical, sensory and motor traits; the growth of manual habits, language habits and intelligence in a superior environment. Data should be obtained from matched individuals in control groups developing under different environmental conditions. The experiment would throw light on the problem of possibilities in nurtural modification of basic traits'.[166] Learning processes had to be studied and improved, children's sleeping habits controlled, their appetites corrected, their 'maladaptive emotional traits' erased, their social skills reinforced and their play habits observed. Child and Ritter maintained that improvement of the race lay not in reproduction but in social environment; such a position was perfectly in agreement with their scientific views on heredity and development.

Child went even further than Ritter. In 1924 he published *Physiological Foundations of Behavior*, probably the wildest and most political of his books. After providing a general reassessment and recapitulation of his gradient theory, in the last chapters he readily applied his notion of dominance and subordination to society and its modes of organization. As he had previously argued in 1921, the organism was not a well-defined thing or a complex entity; it was a process, a timely thing, a behavioural pattern: 'the behaviour of a specific protoplasm in a certain environment'.[167] To Child, such a conception of organism could effectively represent and be extended to the 'social organism' itself:

> The sociologist is accustomed to call the relatively fixed and permanent social patterns institutions. Are there not patterns or orders within the organism fundamentally similar to social institutions in that they represent integrations of living units? Is not the organism itself and institution of some sort in the living protoplasm of one or more cells? If the physiological conception of axiate pattern developed in earlier chapters and elsewhere is correct the axiate organism is an institution resembling the state. It represents fundamentally a relation of dominance and subordination, i.e., of government and governed.[168]

If the 'animal' organism and the 'social' organism were subject to the same dynamics and laws – gradients, integration, dominance and subordination – then, as Ritter and Scripps had argued years before, biology had to be seen as the true foundation of sociology. In other words, sociological disciplines had to be based on what Child dubbed 'sociological biology'. Child used the work of his younger colleague at the University of Chicago, the ecologist and activist W. C. Allee (1885–1955), to prove the consistency of the analogy between organism and society. Allee had studied diverse phenomena of animal aggregations and, with Espinas and Kropotkin, had concluded that cooperation, not competition, dominated the evolution and behaviour of animal communities (including humans) – a belief that that eventually brought him to support some pacifist and socialist options during and between the two world wars.[169]

Although Child did not share Allee's optimism about animal and human cooperation, he embraced Allee's understanding of the relations between biology, ecology and its bearings on human society. In particular, he embraced the idea that the behaviour of an individual organism could depend on its dynamic relation with the whole group. Furthermore, the reciprocal adjustment, instinctive or intelligent, of each member to the needs of the community made social integration possible. Just as single cells were united and integrated in a unique whole (what Child called 'organismic integration'), single animals were integrated in social wholes (namely, 'social integration'). However, to Child, social integration was not something given or primordial, it was something built according to specific rules and laws; it was a product of organic and social evolution. Just as in the simplest forms organismic integration is partial and limited, social integration in primitive societies – tribes or clans – is very unstable and variable.[170] Now, higher organic forms demonstrated a progress, an advancement, an evolution towards integration: the dominant region better controlled subordinate regions so that cellular order followed an axiate pattern. Likewise, from primitive societies to modern states, better forms of social organization developed, forms assuring better types of integration – namely, better types of dominance and power upon the behaviour of subordinated individuals. From anarchy to monarchy, from oligarchy to democracy, dominance progressively changed. In modern democratic societies, Child explained, 'the state possesses social polarity and may be termed an axiate social pattern, with government representing the dominant region and the various classes and differentiations of its members in relation to the government the subordinate regions of various levels of the axis'.[171]

In fact, with Ritter, Child's organicism was not supporting a totalitarian regime. He believed that communism or any totalitarian form of organization was only a primitive phase towards better stages of social integration:

> The progress of evolution then has apparently not been toward a socialistic or communistic form of integration. On the contrary, those types of integration which approach communism most closely, *e.g.*, the colonial animals such as the sponges, hydroids, etc. and the multiaxial plants have been left far behind in the course of evolutionary progress and such success as they have attained appears to be due to the fact that they are not strictly communistic but possess some degree of autocratic or oligarchic dominance.[172]

Organic evolution proved that the best type of physiological integration relied on a certain kind of democratic organization. To Child, the vertebrate body was not an autocracy of cells but a democracy of organs conspiring for the welfare of the whole. With Ritter and Scripps, Child concluded that after billions of years, the whole evolutionary process proved, both in the organic and the social world, that democracy represented the best type of organization – an organization based on American values. Such an evolutionary and teleological view of history reminds us of the old idealist proposals of Hegel.

A state based on social hierarchies was therefore not interpreted as a despotic organization run by absolute power and dominance, but as a form of harmonic order run by a representative government. The subordinated classes were dominated because they wanted to be. In a democratic state, dominance (power) comes not from a particular person or privileged group but from an idea: the idea of nation or state. In other words, the individual is subordinated to an idea guiding the whole community. He is profoundly influenced by the community's values and ideas; all his actions, needs and activities are essentially shaped by it, because what really makes a person is not his heredity, his germ plasm or genes, but his social relations: 'it is sufficiently obvious that the human individual is not simply the product of heredity. His social relations, or more specifically, his subordination in some way and in some degree to more or less definite ideas play a fundamental part in determining the course of his thought, his sentiments and his behavior'.[173] This is precisely what Ritter was trying to demonstrate at Scripps with his project on child development: that the individual was essentially the product of his social environment and not merely a biological outcome of his parents.

# CONCLUSION: WHATEVER HAPPENED TO ORGANISMAL BIOLOGIES?

In this study we have seen that a small international community of scientists were committed to formulating an anti-mechanistic, anti-reductionist, holistic biology.[1] We have learned that their approach and ideas were not a result of a new paradigm emerging in the twentieth century, but were a complex reformulation or re-appropriation of an old tradition whose original source was Kant and the subsequent Romantic tradition. In addition, we have seen that such a reformulation underpinned an alternative form of biology, an alternative to the gene-centric views of heredity, development and evolution that were gaining momentum in the 1930s with the synthesis of neo-Darwinism and Mendelism. Such an alternative view, with all its variations and idiosyncrasies, saw developmental biology as the privileged science for the understanding of speciation. Evolution could not be conceived as change in the gene frequency in populations, but as systemic changes happening during ontogeny. We have also found that this tradition, as transplanted and translated in England and the United States, acquired a specific and contingent political meaning and force. The notion of the organism as an irreducible system of interdependent causes and effects worked as a powerful political metaphor which could inspire a more equitable, harmonious and efficient society. At the same time, mechanistic and reductionist approaches, exemplified in Mendelism and Weismannism, came to be associated with capitalist ideology and a conception of society as comprising an aggregate of selfish individuals in constant competition.

At its height, the Romantic bio-philosophical tradition reconstructed here commanded the allegiances of some outstanding scientists working at prestigious institutions. But it is a tradition in need of reconstruction precisely because it subsequently went into decline, experiencing revivals thereafter. Probably one of the most interesting and eloquent examples of the tradition's decline is the Scripps Marine Association. From as early as 1912 – when the Scripps Association officially became part of the University of California, Berkeley and was renamed the Scripps Institution – things began to change. From that time, the unconventional agenda of the institution – its emphasis on the organism as a whole, its social and political programme based on scientific advice, its eccentric educational plans – came under fire from the University of California faculty. B. Kovarik describes how Scripps himself reacted against academic obstructions and criticism, accusing the academics of destroying his creation. In writing to Ritter in 1915, he complained that 'a bunch of wooden-headed visionless university men ... burned down our temple to roast a little pig'.[2] Even though Ritter was able to maintain a certain 'unconventionality' during his directorship, after his retirement in 1923, the Scripps Institution of Oceanography became a more 'conventional' place; the strong emphasis on organismal biology which had characterized the institution during its early years faded away.

The reasons for the eclipse of the Scripps agenda and of organicist views more generally are certainly complex. From a wider perspective, Gilbert and Sarkar have argued that there were two important reasons for the decline of organism: the association with vitalist proposals – such as those of Driesch and von Uexkull – and the link with totalitarian philosophies.[3] Both Nazi and Communist ideologies undermined the scientific credibility of the organicist sciences. This is true, but it is not enough. There were many interconnections among philosophical, political, institutional and scientific spheres that emerge vividly through specific cases, institutions and scientists. Indeed, not very far from Ritter's lab in southern California, a new, powerful and well-founded conceptual paradigm was emerging: what Lily Kay dubbed the 'protein paradigm'. From the 1930s onwards, the Rockefeller Foundation began supporting, with huge amounts of funding, a new 'physical biology' at the California Institute of Technology (Caltech) in Pasadena: what in 1938 Warren Weaver redefined as the molecular biology programme. As Kay explains, molecular biology:

> aimed to discover general psychochemical laws governing vital phenomena. In so doing it distanced its concerns from emergent properties, from interactive processes occurring within higher organisms, between organisms (e.g. symbiosis), and between organisms and their environment, thus bracketing out of biological discourse a broad range of phenomena generally subsumed under the term 'life' ... the new biology generally acknowledged only mechanisms of upward causation, ignoring the explanatory role of downward causation.[4]

Such a new biology, in employing the methods of physicists and chemists, was essentially reductionist, mechanistic and averse to old biological traditions: 'Caltech's biology program explicitly aimed to depart from established biological tradition in order to create a new science of life based on cooperation with the physical science and engineering'.[5] Yet the new physical biology was essentially laboratory-focused and unfavourable to field studies: 'Amplifying and formalizing the dominant trend of modern science, animate nature within the new program increasingly retreated from the field into the confines of [the] laboratory'.[6] Needless to say, the new molecular paradigm was hugely successful in convincing, attracting and diverting many young brilliant scientists from the organismal field. Indeed, the Rockefeller Foundation, in agreement with California 'crackpotism' and, therefore, with the political utopianism of the Scripps Association, persuasively linked scientific research with the technological improvement of the nation, and scientific bio-knowledge with the promise of the concrete physiological improvement of the human race. Indeed, as Kay noted, the molecular biology programme was inscribed in the foundation's 'Science of Man' agenda. For the Rockefeller Foundation, supporting life science meant fostering and nurturing, at the same time, human sciences broadly conceived. The prestige of physical and chemical sciences guaranteed and promised, much more than the anti-mechanistic objections and holistic prescriptions of Ritter and Scripps, rigorous explanations and effective predictions in the biomedical and human sciences. The Rockefeller Foundation, like the Scripps Association, connected science and utopia. But the foundation, thanks to Caltech's people, did it in a much more convincing way. As Kay concluded:

> A biology governed by faith in technology and in the ultimate power of upward causation is far more amenable to strategies of control than a science of downward causation, where elements cannot be fully understood apart from the whole. There is seductive empowerment in a scientific ideology in which the complexities of the highest levels can be fully controlled by mastering the simplicity of the lowest.[7]

The trend towards the molecularization of life was not merely confined to Caltech, however. Successful results backed and reinforced the protein paradigm in diverse fields. Morgan's chromosome theory of heredity had been extremely successful, and after the 1930s it was widely accepted in the scientific community. The decisive views of Morgan's ambitious students at Columbia University – figures such as A. Sturtevant, C. Bridges and H. J. Müller, and then T. Dobzhansky – on the way that heredity should be studied overshadowed other approaches and methods. Morgan's convincing separation between transmission and expression, genetics and embryology, provided a new model of how heredity should be studied. Heredity was now effectively defined in terms of genetics.[8] In addition, the triumph of genetics opened the door wide for research on the physical basis of heredity, culminating with Watson and Crick's discovery of DNA's double-helical structure in 1953. The progressive molecularization of the gene was unfavourable to organismic ideas. As the Canadian developmental biologist B. Goodwin (a former student of Waddington) observed in 1994: 'Organisms, those familiar plants and animals, including ourselves, that we see all about us, as well as the many invisible forms such as bacteria and other microbes, have disappeared as the fundamental units of life. In their place we now have genes, which have taken over all the basic properties that used to characterize living organisms'.[9] The triumph of molecular biology after World War II left little space for the organismal conception of biology.[10] Successful theories do not only spread in the scientific community; they also convey methods, practices and selected phenomena for study. Ritter's developmental agenda was totally at odds with the triumphant new theories, and the epistemology underlying them. The developmental programme, together with its holistic understanding of biological phenomena, faded slowly away from the mainstream interests of the scientific community. In fact, from 1930 onwards development increasingly ceased to be the central phenomenon to investigate in order to understand heredity and evolution.[11] Evolution and heredity could now be explained by gene transmission.

The impact of the protein paradigm was not only in opposition to the Scripps agenda. Things went similarly at the University of Chicago. Mitman has shown how, after the retirements of Lillie, Child and the old guard, and especially after World War II, organismic biology, with its emphasis on cooperation, self-organization and openness to Lamarckism, became politically uncomfortable and therefore lost popularity among younger biologists. Regarding ecological sciences, Mitman noted that 'the organicist, cooperative renditions of nature that were the hallmark of Chicago ecology became increasingly difficult to sustain in the political climate of the cold war period in which group conflict and competition were seen as essential to a pluralistic, democratic society. A new ideological foundation for biological humanism was in order'.[12] The disturbing similarities between organismic philosophical views about heredity and Lysenko's theory of environmentally acquired characters did not help the organismal cause. In fact, the harsh critiques that had been issued against Mendelian genetics before World War II could no longer be reiterated: they were dangerously similar to what Lysenko was saying.[13]

A further reason for the decline should be mentioned: the influential and persuasive views of the advocates and followers of the modern evolutionary synthesis. After the 1930s the claim that evolutionary novelties could be linked to small, chance genetic mutations, adaptively and gradually accumulated by natural selection, acquired extraordinary force. For some, Mendelian genetics combined with Darwinian evolution seemed to clarify all that biology had to explain. As with Morgan in the 1930s, developmental concerns could be left out once again from the Darwinian core. As D. Depew and B. Weber recorded: 'The increasingly successful integration of genetics with evolutionary theory left developmental biologists more isolated than they had been before. Developmental biology played virtually no role in the formation of the modern synthesis'.[14] Although the modern synthesis was just one element within a much wider, complex, articulated and interesting context,[15] it cannot be denied that Darwinian evolutionary thought, with its gradualist and populationist emphasis, was at home with the emerging gene-centric paradigms. Genetic Darwinism and molecular biology, Depew and Weber concluded, 'were going from strength to strength'.[16] In other words, a certain popular narrative of the modern synthesis was made more and more consistent with the epistemology of newly emerging and successful disciplines such as molecular biology, genetics and biochemistry.[17] Most of the organismal biologists would never have accepted the philosophical basis of these disciplines, just as they had never accepted Weismann's neo-Darwinism.

## Reinventing the Wheel?[18]

Nevertheless, organismic biology survived; Romantic bio-philosophy is still among us, albeit in new guises. Indeed, a few individuals continued to support forms of organismic biology after World War II. Woodger lived until 1981, but he died quite unknown and with no influential pupils or school. Others tried to adapt more or less holistic views to Mendelian genetics. In the 1950s and 1960s figures such as C. H. Waddington, R. B. Goldschmidt (1878–1958), I. I. Schmalhausen (1884–1963) and L. von Bertalanffy, with their epigenetic and systemic thinking, revived some of the ideas that had characterized the old organicist school. Critical of gene-centric, reductionist and mechanist approaches to life sciences, they aimed for an effective integration of development and evolution, genetics and embryology, within an organicist framework. However, and notwithstanding their authority and influence, they had no relevant impact and stimulated no change within the scientific world.

Arguably, it was only during the late 1970s that there gradually emerged a new sympathy towards organismic themes. A former pupil of Dobzhansky's, the evolutionary geneticist Richard Lewontin, began publishing articles and books challenging, and then arguing vociferously against, gene-centric, deterministic and reductionist views and approaches to biology and evolution. As Lewontin, Rose and Kamin explained in 1984, in briefly outlining their dialectical model of explanation in biology: 'For dialectics the universe is unitary but always in change; the phenomena we can see at any instant are parts of processes, processes with histories and futures whose paths are not uniquely determined by their constituent units. Wholes are composed of units whose properties may be described, but the interaction of these units in the construction of the wholes generates complexities that result in products qualitatively different from the component parts'.[19] Ritter or Child could not have expressed it better. In addition, in the same years other biologists and philosophers followed the same path. S. Oyama, D. Noble, S. F. Gilbert, S. C. Stearns, L. Buss, A. Wallace, B. Goodwin, W. M. Elsasser and, later, E. F. Keller, R. Rosen, C. N. van der Weele, J. Dupré, W. Callebaut and G. Müller, together with many other scholars, began to reintroduce organismic discourses.[20] Cumulatively, these have had an impact. The 'organism' returned, with all its metaphorical power.

Oyama's successful book *The Ontogeny of Information* (1985) opened a harsh debate in biology and psychology. In her account, organisms had to be seen not as genes' machines, as complex entities guided by small God-like molecules – the genes – containing all the necessary information for driving development and behaviour. Instead, the organism is a system in constant and dynamic interaction with the environment. As Ritter had claimed, a democracy of causes had to be invoked, and the genes had to be seen as mere systemic 'interactants' among many others. As she argued, the new biology is moving towards 'a conception of developmental system, not as reading off of a preexisting code, but as a complex of interacting influences, some inside the organism's skin, some external to it, and including its ecological niche in all its spatial and temporal aspects'.[21] The form of an organism is not built by genes, but by developmental processes where networks of causes, at different levels of biological organization, interact. For Oyama, one of the major difficulties in accepting a systemic view of life lies in the heritage of our thinking, which sees matter as essentially inert. The consequence of such an assumption is that matter, in order to be organized and form a living system, needs an active agent: a soul, a vital force or information encoded in the genes. But such a metaphysical assumption does not need to be accepted. Together with Kant, Hegel, Schelling, Whitehead and many of her organicist predecessors, Oyama argued that matter is indeed reactive, active and productive: 'Matter, including living matter, is inherently reactive, and change, far from being an intrusion into some static natural order, is inevitable'.[22]

In agreement with many nineteenth- and twentieth-century Romantic naturalists – who as we saw were critical of the mechanistic philosophies of the eighteenth century – Oyama intended to transcend old dichotomies such as matter and form, nature and nurture. With Woodger, she believed that the dichotomy between innate and acquired had to be reconfigured. Indeed, within the scheme of Oyama's developmental systems theory (DST), what is inherited is not fixed at the beginning; it progressively emerges during ontogeny. Woodger distinguished between intrinsic and relational properties in heredity. Intrinsic properties were old relational properties fixed during previous ontogenetic cycles; variations between intrinsic and relational properties were related to variations in ontogeny and phylogeny. For Oyama, what are inherited are integrated developmental systems. Changes in these systems bring about changes in the organismal form, and therefore in evolution. Indeed, along with Child and Ritter, Oyama saw evolution as the contingent change of developmental systems (as we saw, the old organicists used the expression 'systems of reaction'). Like most of the figures mentioned in this book, Oyama thought that natural selection worked as a filter at most. It could not create anything, only prune the unfit. The creative side of speciation relied on ontogeny. As she explained, echoing Ritter's 'biotic genesis', 'Evolution works not only on alternative finished phenotypes ...

but on alternative ontogenetic pathways as well, at organismic and intraorganismic levels'.[23] Finally, Oyama also shared with the old organismal biologists a concern for the social consequences of a deterministic and reductionist science of life. A science instantiated by a gene-centric view of development and evolution not only is ideological and unintelligible; it is politically dangerous because it supports and legitimizes a limited and conservative vision of human nature.

In the 1990s and 2000s other critiques of genetic determinism and reductionism in biology emerged. As the Harvard biologist R. Hubbard claimed in 1993: 'The fact is that DNA doesn't "do" anything; it is a remarkably inert molecule. It just sits in our cells and waits for other molecules to interact with it'.[24] A few years later a physicist turned molecular biologist, E. F. Keller, also argued that genes alone are powerless. The DNA molecule was inert without the entire cellular machinery working around it: 'the gene can no longer be set above and apart from the processes that specify cellular and intercellular organization. That gene is itself part and parcel of processes defined and brought into existence by the action of a complex self-regulating dynamical system in which, and for which, the inherited DNA provides the crucial and absolutely indispensable raw material, but no more than that'.[25] Similar arguments proliferated. 'Genes are certainly not "self-replicating"', claimed Lewontin, as 'New strands are synthesized by a complex cell machinery consisting of proteins that use the old strands as templates'.[26] 'What does an organism inherit?', Griffiths and Gray asked; 'Certainly more than the nuclear DNA', they concluded one year later.[27] Old metaphors such as the genetic blueprint, programme, recipe or code needed to be replaced by interactive networks, cycles of contingencies and developmental processes. There was no 'blueprint' or 'code' beyond the complex organismic system implementing it. The whole cell and its environment had to be taken into account. By the end of the 1990s, thanks to the wide success of a relatively novel discipline dubbed 'evo-devo', convincing new organismic models of development and evolution had been proposed. In 2003 two biologists, G. B. Müller and S. A. Newman, aimed for a new direction for evolutionary biology, a direction centred not on genes but on the organism and its ontogeny. The old neo-Darwinian, gene-centred paradigm, they reminded readers, could not explain the origination of the organismal form because natural selection 'has not innovative capacity: it eliminates or maintains what exists. The generative and ordering aspect of morphological evolution are thus absent from evolutionary theory'.[28] Evolutionary biology was much more than evolutionary genetics, they stressed. Epigenetical factors played a much more important role than previously expected for the creative and generative side of evolution. Speciation is not only about gene mutation and selection; it is also about developmental variation arising from the dynamic organism – environment interaction.

Complaints about the overlooked role of the environment in ontogeny began to be widespread. In 1999 the Dutch philosopher of biology van der Weele published a book describing and discussing the environmental causes of development. She asked why the environment was so absent in textbooks of developmental biology. A big part of the answer was that 'the dominant approach to development is genetic'.[29] In assuming that development is nothing but the expression of a genetic programme, many developmental biologists simply ignored, or basically overlooked, the fact that such expression always happens in an environmental context. Yet in spite of the general assumption according to which the environment is mere background or noise, there are many known observations showing that the environment is, in many cases, causally determinant in ontogeny. From the aphid's development to the butterfly's seasonal polyphenisms, from the sexual determination of echiurid to the predator-induced polyphenisms in *Daphnia cucullata* - all show that the environment is much more important than commonly believed (and mentioned in textbooks). Once again the main message was that the organism is not an assemblage of genes but a whole dynamic system in interaction with its own environment.

The well-known developmental biologist Scott F. Gilbert is also part of such an international resurgence of organicist thinking. An article published in 2000 with S. Sarkar provides a brief history of organicism. From Kant's bio-philosophy to nineteenth-century experimental embryology, organicism provided an ideal theoretical framework for the understanding of organism's plasticity. After a period of decline during the central decades of the twentieth century, the organicist view is emerging again. In going beyond orthodox gene-centrism, the organicist stance is informing contemporary developmental biology. As Gilbert and Sarkar claim: 'Many of the principles of organicism remain in contemporary developmental biology, but they are rarely defined as such ... Based on principles or organicism, developmental biology should become a science of emerging complexity'.[30]

Developmental biology is not the only contemporary discipline to be inspired by organicism. Evolutionary developmental biology (evo-devo) is another recent discipline that uses, or mentions implicitly, many of the ideas that characterized the old organicist approach. As B. Hall recalls, evo-devo has its origins in the evolutionary embryology of the nineteenth century, but it only began thriving in the 1980s.[31] Gilbert defines evo-devo as the science seeking to 'discern how the mechanisms of development effect evolutionary change and stasis, how changes in development generate evolutionary novelty, how development might constrain certain phenotypes from arising, and how developmental mechanisms themselves evolve'.[32] Many components of this research programme are reminiscent of the agendas of Child, Lillie, Russell or D'Arcy Thompson. An example is the deep interest in the relation between growth, form and speciation. In fact, in perfect agreement with Thompson's old agenda, the American biologist Richtsmeier argues: 'The process of growth offers the potential for evolutionary change through the range of diversity in, and interaction between, elements of timing, direction, and magnitude of local change in growth patterns of individuals'.[33] The evo-devo organismic model that W. Callebaut, G. B. Müller and S. A. Newman presented in 2007 extends and synthetizes all these insights. As they affirm, 'The organismic system approach (OSA) argues that the majority of novelties [in evolution] result from higher-level organizational properties, epigenetic interactions, and environmental influences that arise during the modification of established developmental systems. In this sense, novelties would be epigenetic by-products of the system properties of development'.[34] Most of the figures mentioned in this book would have easily approved of such a proposal.

In the very last few years biologists, bio-philosophers and historians of science have begun to revise and extend our knowledge about the old modern synthesis. The conviction is now widespread that the modern synthesis failed to include important disciplines such as developmental biology and ecology.[35] Some historians and philosophers of science have remarked that the modern synthesis was far from a coherent and 'synthetic' movement; others have observed that the synthesis historiography was too focused on Anglo-American scholarship.[36] The purpose of the recently developed extended synthesis (ES) is to obviate such limitations. As a consequence, developmental biology and ecology are now fully included in a larger and more pluralist framework. This means that other traditions and scholarships are also being considered, and historians of science are beginning to uncover some neglected authors and ideas outside the Anglo-American context.[37] In 1987 L. Buss got it right when he commented that the modern synthesis could not have a theory of individuality precisely because individuality was assumed; ontogeny was accepted as the unproblematic background and not as an issue to be investigated. As he concluded almost thirty years ago: 'It may seem foolhardy, in today's climate, to advocate anything but the genic selection view. The lion's share of research funds is directed at molecular approaches to biological problems'.[38] In light of these observations, it is not surprising that people such as Child, Ritter and other members of the group of biologists I discussed in this book were mostly forgotten. They could never be part of the modern synthesis agenda because, for them, ontogeny – the organismal form and its systemic variations – was the central problem of biology. They exemplified and supported an *alternative synthesis*.

However, organicist approaches and views are becoming an integral part of the new synthesis of evolutionary biology and other disciplines. In the most recent findings in genomics, metagenomics, embryogenesis, oncogenesis and ecology, we can see that scientists have started to reintroduce notions such as 'process', 'network', 'emergence', 'system' and 'field'.[39] Might the protagonists of the story told in this book turn out to be the pioneers, precursors and originators of a new 'way of seeing' emerging in biology today? Or will they continue to be seen as mere exponents of a deceased, outmoded tradition, having only historical interest? In his 2005 paper 'Other Histories, Other Biologies', Gregory Radick posed two important questions about biological science: 'Would quite different histories have produced roughly the same science? Or, on the contrary, would different histories have produced other, quite different biologies?'[40] In light of the history of the Romantic bio-philosophical tradition, I would be inclined to answer the second question in the affirmative. In all probability, if the organismal biologists of the late nineteenth and early twentieth centuries had succeeded, a different biology would have been developed – a biology where the heroes would have been neither Weismann and Mendel nor Bateson and Morgan, but Wolff, Kant, Blumenbach, von Baer, Goethe, Cuvier, Barry and Owen, followed by such twentieth-century figures as Whitman, Child, Russell, D'Arcy Thompson and many others I have discussed in this study.

# NOTES

## Introduction

1. J. W. Goethe, *Faust* (New York: Bantam Classics, 1988), Act I, scene iv.
2. R. Safranski, *Romantik. Eine deutsche Affäre* (Munich: Hanser, 2007), trans. M. Ritterson, 2008, *Goethe Institute, German Literature Online*, at http://www.litrix.de/mmo/priv/24106-web.pdf [accessed 9 August 2013].
3. As Barlow notes, the aphorism conveyed the idea that 'the properties of the cells are fashioned, or determined, according to their context within the developing whole; it is not the cells that fashion the plant, or direct its morphogenesis, for this is a phenomenon that transcends the individual cell'; P. W. Barlow, '"The Plant Forms Cells, Not Cells the Plant": The Origin of de Bary's Aphorism', *Annals of Biology*, 49:2 (1982), pp. 269–72, on p. 270.
4. A. Reynolds, 'The Cell's Journey: From Metaphorical to Literal Factory', *Endeavour*, 31:2 (2007), pp. 65–70.
5. C. O. Whitman, 'The Inadequacy of the Cellular Theory of Development', *Journal of Morphology*, 8:3 (1893), pp. 639–59, on p. 655.
6. Ibid., p. 653.
7. See A. Cunningham and N. Jardine (eds), *Romanticism and the Sciences* (Cambridge: Cambridge University Press, 1990); and S. Poggi and M. Bossi (eds), *Romanticism in Science: Science in Europe, 1790–1840* (Dordrecht: Kluwer Academic, 1994).
8. J. Reddick, 'The Shattered Whole: Georg Buchner and Naturphilosophie', in A. Cunningham and N. Jardine (eds), *Romanticism and the Sciences* (Cambridge: Cambridge University Press, 1990), pp. 322–40, on p. 336.
9. Poggi and Bossi (eds), *Romanticism in Science*, p. xv.
10. See F. C. Beiser, *German Idealism: The Struggle against Subjectivism 1781–1801* (Cambridge, MA: Harvard University Press, 2002).
11. L. W. von Bertalanffy, R. Ashby and G. M. Weinberg, *Trends in General System Theory* (Hoboken, NJ: John Wiley & Sons, 1972).
12. M. Beckner, 'Organismic Biology', in P. Edwards (ed.), *Encyclopedia of Philosophy* (London: Macmillan, 1969), p. 549.
13. E. Mayr, 'The Autonomy of Biology: The Position of Biology among the Sciences', *Quarterly Review of Biology*, 71:1 (1996), pp. 97–106, on p. 99.
14. D. Haraway, *Crystals, Fabrics, and Fields* (Baltimore, MD: Johns Hopkins University Press, 1976), p. 2.
15. F. S. Gilbert and S. Sarkar, 'Embracing Complexity: Organicism for the 21st Century', *Developmental Dynamics*, 219:1 (2000), pp. 1–9.

16. See E. Ungerer, *Die Teleologie Kants und ihre Bedeutung für die Logik der Biologie* (Berlin: Verlag von Gebrüder Borntraeger, 1922).
17. E. Cassirer, *The Problem of Knowledge: Philosophy, Science and History since Hegel* (New Haven, CT: Yale University Press, 1950), p. 214.
18. It is interesting to mention that the philosopher David Lamb connected Hegel's holistic and organicist philosophy with von Bertalannfy's system theory; see D. Lamb, *Hegel: From Foundation to System* (The Hague: Martinus Nijhoff, 1981), ch. 1.
19. A. Harrington, *Reenchanted Science. Holism in German Culture from Wilhelm II to Hitler* (Princeton, NJ: Princeton University Press, 1996), p. 177.
20. For a general introduction to Kant and Romantic bio-philosophy, see the following sections. For a general introduction to the debate about mechanism versus holism and organicism, see G. E. Allen, 'Mechanism, Vitalism and Organicism in Late Nineteenth and Twentieth-Century Biology: The Importance of Historical Context', *Studies in History and Philosophy of Science Part C: Studies in History and Philosophy of Biological and Biomedical Sciences*, 36:2 (2005), pp. 261–83.
21. W. Garstand, 'The Theory of Recapitulation: A Critical Re-statement of the Biogenetic Law', *Linnean Journal-Zoology*, 35:232 (1922), pp. 81–101, on p. 81; see also B. K. Hall, 'Balfour, Garstang and de Beer: The First Century of Evolutionary Embryology', *American Zoologist*, 40:5 (2000), pp. 718–28. On the relation between ontogeny and phylogeny, see also the classic 1977 book by S. J. Gould, *Ontogeny and Phylogeny* (Cambridge, MA: Harvard University Press, 1977).
22. R. Lickliter and T. D. Berry, 'The Phylogeny Fallacy: Developmental Psychology's Misapplication of Evolutionary Theory', *Developmental Review*, 10 (1990), pp. 322–38, on p. 357.
23. R. C. Lewontin and R. Levins, *The Dialectical Biologist* (Cambridge, MA: Harvard University Press, 1985).
24. A. Lovejoy, 'On the Discrimination of Romanticisms', *Publications of the Modern Language Association of America*, 39 (1924), pp. 229–53, on p. 232.
25. Ibid., p. 236.
26. Quoted in I. Berlin, *The Roots of Romanticism* (Princeton, NJ: Princeton University Press, 1999), p. 24.
27. See K. J. Fink, *Goethe's History of Science* (Cambridge: Cambridge University Press, 1991).
28. Laubichler and Maienschein had done very important work in charting the alternative routes of biology during the twentieth century. See M. Laubichler and J. Maienschein, *From Embryology to Evo-Devo: A History of Developmental Evolution* (Cambridge, MA: MIT Press, 2007).
29. K. Schwenk, D. K. Padilla, G. S. Bakken and R. J. Full, 'Grand Challenges in Organismal Biology', *Integrative and Comparative Biology*, 49:1 (2009), pp. 7–14, on p. 7.

## 1 Old and New Organicisms

1. M. Planck, 'Letter', *Nature*, 127:3207 (1931), p. 612. Planck was explaining that with the introduction of quantum mechanics in physics, physical sciences were doomed to change radically. The notion of wholeness had to be introduced. This quotation was often employed enthusiastically by many organismic biologists (in particular J. S. Haldane and J. H. Woodger).
2. See Chapter 6 in this volume.

3. G. Werskey, *The Visible College* (London: Penguin, 1978).
4. Yoffe, as Werskey recalls, was one of the most prominent and influential physicists in Soviet Russia; see ibid., p. 138.
5. W. E. Ritter, L. L. Whyte, L. Hogben, J. S. Haldane, H. J. Woodger, E. S. Russell, A. Yoffe and J. Needham, 'Historical and Contemporary Relationships of the Physical and Biological Sciences', Second International Congress on the History of Science (Third Session), *Archeion*, 14:4 (1933), pp. 497–502.
6. Whyte, in ibid., pp. 510–11.
7. Haldane, in ibid., p. 502.
8. Russell, in ibid., p. 506.
9. Woodger, in ibid., p. 514.
10. Needham, in ibid., p. 509.
11. Ibid.
12. Yoffe, in ibid., p. 514.
13. Hogben, in ibid., p. 511.
14. Planck, 'Letter', p. 612.
15. As Whyte clearly argued: 'classical methods are essentially inadequate to deal with ordered structures, and an additional law of a new type is necessary. This new law is expressed by equations called the "Quantum conditions", whose importance for this discussion lies in the fact that they refer directly to systems as a whole, and not to the individual parts or particles which make up the system. For example, classical theory could not account either for the stability or for any of the unitary characteristics of the set of electrons which constitutes a copper atom. But in the quantum theory the quantum conditions for a complex atom imply the recognition of the atom as a system with various definable characteristics possessed by the atom as a whole ... Thus certain aspects of the conflict between the purely micro-analytical methods of classical physics and the organic concepts of biology have been eliminated at their root' (Whyte, in 'Historical and Contemporary Relationships', p. 510).
16. J. S. Haldane, *The Philosophical Basis of Biology* (London: Hodder and Stoughton, 1931), p. 33.
17. See A. Meyer-Abich, *Biologie der Goethezeit* (Stuttgart: Hippokrates Verlag, 1949).
18. J. Needham, 'Organicism in Biology', *Journal of Philosophical Studies*, 3:9 (1928), pp. 29–40, on p. 29.
19. E. S. Russell, *The Interpretation of Development and Heredity* (Oxford: Clarendon Press, 1930), p. 173.
20. For an introduction, see J. H. Zammito, *The Genesis of Kant's Critique of Judgment* (Chicago, IL: University of Chicago Press, 1992).
21. See, in particular, sections 65, 66, 78, 80 and 81 of Kant's *Critique of Judgment*. On Kant's bio-philosophy there is a large amount of literature. In particular, see J. MacFarland, *Kant's Concept of Teleology* (Edinburgh: Edinburgh University Press, 1970); P. McLaughlin, *Kant's Critique of Teleology in Biological Explanation, Antinomy and Teleology* (New York: Edwin Mellen Press, 1990); R. E. Butts, 'Teleology and Scientific Method in Kant's Critique of Judgment', *Nous*, 24:1 (1990), pp. 1–16; H. Ginsborg, 'Kant on Understanding Organisms as Natural Purposes', in E. Watkins (ed.), *Kant and the Sciences* (Oxford: Oxford University Press, 2001), pp. 231–59; and P. Guyer, 'Organisms and the Unity of Science', in E. Watkins (ed.), *Kant and the Sciences* (Oxford: Oxford University Press, 2001), pp. 259–82. See also the excellent introductions to Kant's philosophy of biology in T. Lenoir, *The Strategy of Life: Teleology and Mechanics in Nineteenth-Century German Biology* (Chicago, IL: University of Chicago Press,

1982), and R. J. Richards, *The Tragic Sense of Life: Ernst Haeckel and the Struggle over Evolutionary Thought* (Chicago, IL: University of Chicago Press, 2008).

22. I. Kant, *Kritik of Judgment*, trans. and intro. J. H. Bernard, rev. 2nd edn (London: Macmillan, 1914).
23. Ibid., p. 278.
24. Ibid., p. 280.
25. Ibid., p. 331.
26. R. J. Richards, 'Goethe's Use of Kant in the Erotics of Nature', in P. Huneman (ed.), *Understanding Purpose: Kant and the Philosophy of Biology* (Rochester, NY: University of Rochester Press, 2007), pp. 137–48, on p. 146.
27. Kant, *Kritik of Judgment*, p. 352.
28. As Kant pointed out in section 66: 'it may be that in an animal body many parts can be conceived as concretions according to mere mechanical laws (as the hide, the bones, and the hair). And yet the cause which brings together the required matter, modifies it, forms it, and puts it in appropriate place, must always be judged of teleologically: so that everything may be considered as organized, and everything again in a certain relation to the thing itself in an organ' (ibid., p. 282).
29. Kant was deeply impressed by the phenomenon of organic growth. To him, it showed purposive qualities. As he explained: 'the abortions or malformations in growth, where, on account of some chance defects or obstacle, certain parts adopt a completely new formation, so as to preserve the existing growth, and this produce an anomalous creature ... are among the most wonderful properties of the forms of organic life' (ibid., p. 372).
30. As Lenoir recognizes: 'Immanuel Kant regarded Blumenbach as one of the most profound biological theorists of the modern era' (Lenoir, *The Strategy of Life*, p. 18).
31. See O. Temkin, 'German Concepts of Ontogeny and History around 1800', *Bulletin of the History of Medicine*, 24:3 (1950), pp. 227–46.
32. P. Huneman (ed.), *Understanding Purpose: Kant and the Philosophy of Biology* (Rochester, NY: University of Rochester Press, 2007), pp. 61–100.
33. On *Hydra*'s impact in biology, see V. Dawson, *Nature's Enigma: The Problem of the Polyp in the Letters of Bonnet, Trembley, and Reaumur* (Philadelphia, PA: American Philosophical Society, 1987).
34. See also J. Roger, *Les Sciences de la Vie dans la Pensée Français du XVIII Siècle* (Paris: Armand Colin, 1963); E. Gasking, *Investigations into Generation* (Baltimore, MD: Johns Hopkins University Press, 1968); C. E. Dinsmore, *History of Regeneration Research: Milestones in the Evolution of a Science* (Cambridge: Cambridge University Press, 1991); and finally J. Moscoso, 'Experimentos de Regeneración Animal: 1686–1765. Come defender la Preexistencia?', *Dynamis*, 15 (1995), pp. 341–73.
35. Kant, *Kritik of Judgment*, p. 324.
36. Ibid., p. 280.
37. Ibid., p. 278.
38. Ibid., p. 346.
39. 'In the biological realm, say, in the epigenesis of the foetus as described by Blumenbach (whom Kant knew and read), the various early stages make sense only in relation to their final product: that is, we have to conceive the final stage of development, which is an effect of the earlier stages, as if it were also the cause of the earlier stages'; R. J. Richards, *The Meaning of Evolution: The Morphological Construction and Ideological Reconstruction of Darwin's Theory* (Chicago, IL: University of Chicago Press, 1992), p. 23.

40. See P. McLaughlin, 'Kant on Heredity and Adaptation', in S. Müller-Wille and H. J. Rheinberger (eds), *Heredity Produced: At the Crossroads of Biology, Politics, and Culture, 1500–1870* (Cambridge, MA: MIT Press, 2007), pp. 277–91, on p. 281.
41. Kant, quoted in R. Bernasconi and T. L. Lott, *The Idea of Race* (Indianapolis, IN: Hackett Publishing Company, 2000), p. 10.
42. Kant, quoted in A. Lovejoy, 'Kant and Evolution. II', *Popular Science Monthly*, 78 (1911), pp. 36–51, on p. 44.
43. Kant, quoted in MacFarland, *Kant's Concept of Teleology*, p. 57.
44. Ibid., p. 65.
45. See M. Nicolson, 'Alexander von Humboldt and the Geography of Vegetation', in A. Cunningham and N. Jardine (eds), *Romanticism and the Sciences* (Cambridge: Cambridge University Press, 1990), pp. 169–88.
46. As Nicolson concludes: 'Alexander von Humboldt may be seen as both a product of German Romanticism and an important exponent of a Romantic style within natural inquiry' (ibid., p. 183).
47. As G. H. Müller pointed out, 'von Humboldt owed much to Kant'; G. H. Müller, 'Wechselwirkung in the Life and Other Sciences: A Word, New Claims, and a Concept around 1800 ... and Much Later', in S. Bossi and M. Poggi (eds), *Romanticism in Science: Science in Europe, 1790–1840* (Dordrecht: Kluwer Academic, 1994), pp. 1–14, on p. 5. To Müller, one of the most important notion of modern ecology has connections with the old, Romantic notion of *Wechselwirkung*, e.g. interrelation: '*Wechselwirkung* seems to me to be one of the central topoi of romantic science, and the concept is likely to answer some more of the questions about why and how a scientific discipline such as ecology came to develop' (ibid., p. 3).
48. Humboldt, quoted in A. G. Von Aesch, *Natural Science in German Romanticism* (New York: Columbia University Press, 1941), p. 179. Humboldt's book is *Uber die gereizte Muskel-und Nervenfaser* (1797).
49. E. Nordenskiöld, *The History of Biology* (New York: Knopf, 1935), p. 315. As Helferich also reports: 'Humboldt agreed with Kant that a different approach to science was needed, one that could account for the harmony of nature that lay beneath the apparent diversity of the physical world ... In this great sequence of cause and effect, nothing can be considered in isolation'; G. Helferich, *Humboldt's Cosmos: Alexander von Humboldt and the Latin American Journey that Changed the Way We See the World*, Kindle edn (London: Penguin, 2011).
50. Kant, quoted in MacFarland, *Kant's Concept of Teleology*, p. 108.
51. As Friedman and Nordmann remind us: 'It is now a commonplace that the development of modern mathematics, mathematical logic, and the foundations of mathematics can be profitably seen as an evolution "from Kant to Hilbert". It is our conviction, in addition, that the development of modern scientific thought more generally – including the physical sciences, the life sciences, and the relationships between both of these and the mathematical sciences – can also be greatly illuminated when viewed as an evolution from Kant, through Poincaré, to Einstein and the logical empiricists and beyond'; M. Friedman and A. Nordmann (eds), *The Kantian Legacy in Nineteenth-Century Science* (Cambridge, MA: MIT Press, 2006), p. 1.
52. See F. Gregory, 'Kant's Influence on Natural Scientists in the German Romantic Period', in R. P. W. Visser (ed.), *New Trends in the History of Science* (Amsterdam: Rodopi, 1989), pp. 53–66.

53. N. Jardine, *The Scenes of Inquiry: On the Reality of Questions in the Sciences* (Oxford: Oxford University Press, 1991), p. 112.
54. See F. C. Beiser, 'Kant and Naturphilosophie', in M. Friedman and A. Nordmann (eds), *The Kant Legacy in the Nineteenth-Century Science* (Cambridge, MA: MIT Press, 2006), pp. 7–26.
55. See R. Wellek, *Immanuel Kant in England, 1793–1838* (Princeton, NJ: Princeton University Press, 1931).
56. See Beiser, *German Idealism*, esp. ch. 1, part III.
57. R. Owen, *The Hunterian Lectures in Comparative Anatomy, May and June 1837*, intro. and commentary by P. R. Sloan (Chicago, IL: University of Chicago Press, 1992).
58. Lenoir, *The Strategy of Life*, p. 2.
59. 'Naturphilosophie derives from several sources, but its intellectual core grew from Kant's critical philosophy, Schelling's transcendental idealism, and Goethe's developmental morphology' (Richards, *The Meaning of Evolution*, p. 21).
60. See L. K. Nyhart, *Biology Takes Form: Animal Morphology and the German Universities, 1800–1900* (Chicago, IL: University of Chicago Press, 1995).
61. R. J. Richards, *The Romantic Conception of Life: Science and Philosophy in the Age of Goethe* (Chicago, IL: University of Chicago Press, 2002), p. 12.
62. 'The third Critiques furnished the starting point for the romantics' own theories of aesthetics and biological sciences' (ibid., p. 64). As Beiser specified: 'Kant's methodological views – especially his demand for systematic unity and his insistence upon synthetic a priori principles – were also important for some *Naturphilosophers*. It is indeed somewhat ironic to find the Neo-Kantians criticising the *Naturphilosophers* for a priori speculation and system building when so much of their inspiration for these activities came from Kant himself! Even the method of analogy, for which *Naturphilosophers* had been so severely criticized, has its Kantian roots' (Beiser, 'Kant and Naturphilosophie', p. 9).
63. Berlin, *The Roots of Romanticism*, p. 68.
64. Fink, *Goethe's History of Science*, pp. 11.
65. See ibid., pp. 75–7 and p. 139.
66. See G. von Molnar, 'Goethe's Reading of Kant's Critique of Esthetic Judgment', *Eighteenth Century Studies*, 15 (1982), pp. 402–20.
67. T. Lenoir, 'The Eternal Laws of Form', in F. Amrine, F. J. Zucker and H. Wheeler (eds), *Goethe and the Sciences: A Reappraisal* (Dordrecht: D. Reidel, 1987), pp. 17–28, on p. 21. On the relation between Kant's third Critique and Goethe, see also J. Holland, *German Romanticism and Science: The Procreative Poetics of Goethe, Novalis and Ritter* (London: Routledge, 2009), p. 20.
68. On Goethe's teleological views, see also J. F Cornell, 'Faustian Phenomena: Teleology in Goethe's Interpretation of Plants and Animals', *Journal of Medicine and Philosophy*, 15 (1990), pp. 481–92.
69. See R. H. Brady, 'Form and Cause in Goethe's Morphology', in F. Amrine, F. J. Zucker and H. Wheele (eds), *Goethe and the Sciences: A Reappraisal* (Dordrecht: D. Reidel, 1987), pp. 257–300, on p. 289.
70. Goethe, in K. M. Meyer-Abich, 'Self-Knowledge, Freedom and Irony: The Language of Nature in Goethe', in F. Amrine, F. J. Zucker and H. Wheeler (eds), *Goethe and the Sciences: A Reappraisal* (Dordrecht: D. Reidel, 1987), pp. 351–71, on p. 361.
71. See Cassirer, *The Problem of Knowledge*, p. 138. I add that we would never understand Goethe's bio-philosophy if we approach his morphological views through the neat dichotomy between form and function.

72. See T. Appell, *The Cuvier–Geoffrey Debate: French Biology in the Decades before Darwin* (Oxford: Oxford University Press, 1987).
73. S. S. Hahn, *Contradiction in Motion: Hegel's Organic Concept of Life and Value* (Ithaca, NY: Cornell University Press, 2007), p. 11.
74. Ibid.
75. See C. Ferrini, 'The Transition to Organics: Hegel's Idea of Life', in S. Houlgate and M. Baur (eds), *A Companion to Hegel* (Oxford: Wiley-Blackwell, 2011), pp. 203–24.
76. As Beiser also reminds us in his introduction to Hegel: 'For the organic view of the world appears throughout Hegel's system. It plays a fundamental role in his logic, ethics, politics, and aesthetics. Hegel understood all these fields in essentially organic terms. The predominance of the organic concept in Hegel's system derives not least from his naturalism: since everything is part of nature, and since nature is an organism, everything must be shown to be part of the organism'; F. C. Beiser, *Hegel* (London: Routledge, 2005), p. 80.
77. Hahn, *Contradiction in Motion*, p. 27.
78. Ibid., p. 31.
79. Hegel's critique of classical logic must be seen in this context. To Hegel, indeed, classical logic was essentially static; it could never help us to understand the reality, precisely because reality is dynamic and historical. Classical logic is, for Hegel, a pure abstraction. Hegel's *Science of Logic* was, as Beiser argues, a logic of life (Beiser, *Hegel*, p. 81).
80. See D. Seamon and A. Zajonc, *Goethe's Way of Science: A Phenomenology of Nature* (Albany, NY: SUNY Press, 1988).
81. See Hahn, *Contradiction in Motion*, p. 82.
82. Beiser, *Hegel*, p. 82.
83. Ibid., p. 85.
84. The intellectual genealogy behind Hegel's organicist *naturphilosophie* is complex. Beiser mentions, among others, Plato (especially Timaeus) and Spinoza; see ibid.
85. Von Aesch, *Natural Science*, p. 185.
86. See S. Sturdy, 'Biology as Social Theory: John Scott Haldane and Physiological Regulation', *British Journal of the History of Science*, 21 (1988), pp. 315–40; and P. J. Bowler, *Reconciling Science and Religion: The Debate in Early-Twentieth-Century Britain* (Chicago, IL: University of Chicago Press, 2001).
87. See J. L Esposito, *Schelling's Idealism and Philosophy of Nature* (Lewisburg, PA: Bucknell University Press, 1977).
88. W. E. Gerabek, *Friedrich Wilhelm Joseph Schelling und die Medizin der Romantik: Studien zu Schellings Würzburger Periode*, Europäische Hochschulschriften, ser. 7, sect. B, vol. 7 (Frankfurt: Peter Lang, 1995). See also G. B. Risse, 'Kant, Schelling, and the Early Search for a Philosophical "Science" of Medicine in Germany', *Journal of the History of Medicine*, 27:2 (1972), pp. 145–58.
89. Schelling, quoted in I. H. Grant, *Philosophies of Nature after Schelling* (London: Bloomsbury Academic, 2008), p. 12. See also L. Montiel, 'Filosofía de la ciencia médica en el romanticismo alemán. La propuesta de Ignaz Döllinger (1770–1841) para el estudio de la fisiología', *Medicina y Historia*, 70 (1997), pp. 1–15.
90. Beiser, *Hegel*, p. 106.
91. Von Baer, quoted in J. Oppenheimer, *Autobiography of Karl Ernst von Baer* (New York: Science History Publications, 1986), p. 12.
92. Ibid., p. 131.

93. See L. Montiel, 'Más allá de El Nacimiento de la clinica. La comprensión de la Anatomia general de Bichat desde la Naturphilosophie de Schelling', in O. Market and J. R. de Rosales (eds), *El inicio del Idealismo Alemán* (Madrid: Editorial Complutense, 1996), pp. 315–25.
94. Esposito, *Schelling's Idealism*, p. 130.
95. Translation from German. See F. W. J. Schelling, *Werke* (Leipzig: Band 1, 1907).
96. M. L. Heuser, 'Schelling's Concept of Self-Organization', in R. Friedrich and A. Wunderlin (eds), *Evolution of Dynamical Structures in Complex Systems* (Berlin: Springer, 1992), pp. 395–415; F. W. J. Schelling, *Escritos sobre la filosofía de la naturaleza*, ed. A. Leyte (Madrid: Alianza Editorial, 1996); and X. Tilliette, *Schelling: une philosophie en devenir* (Paris: Vrin, 1970).
97. Schelling, quoted in Grant, *Philosophies of Nature*, p. 131.
98. Esposito, *Schelling's Idealism*, p. 237
99. Hall, quoted in ibid., p. 198.
100. Peirce, quoted in ibid., p. 203. Esposito also includes J. Stallo, L. P. Hickok, J. Fiske, W. T. Harris and J. Royce as philosophers influenced by German idealism in the USA.
101. Nyhart, *Biology Takes Form*, p. 8.
102. F. C. Beiser, *The Romantic Imperative: The Concept of Early German Romanticism* (Cambridge, MA: Harvard University Press, 2004), p. 156.
103. For example, the organicism defended by Goethe was probably inspired by Kant, and also by Spinoza.
104. As Zammito rightly reminds us: 'Many of the disciples Kant recruited in Germany simultaneously reverenced Lessing and Goethe, read Herder with attentiveness and appreciation, and found Spinoza and pantheism fascinating. Indeed, one of the crucial facts that must be retrieved from the context is the widespread conviction of the incompleteness of the Kantian system and of the agenda for philosophy which that created. Kant contributed substantially to this sense of the openness of his system and to the idea of its possible completion, and only very late, when it became apparent that what his disciples had made did not suit him, would he give public notice that his own works constituted and altogether complete system, and that his heirs had utterly misunderstood him' (Zammito, *The Genesis of Kant's Critique of Judgment*, p. 14).
105. See P. J. Pauly, 'The Appearance of Academic Biology in Late Nineteenth-Century America', *Journal of the History of Biology*, 17:3 (1984), pp. 369–97.

## 2 Romantic Biology: Establishing Connections in the Nineteenth and Twentieth Centuries

1. G. Cuvier, *Discours sur les révolutionnes de la surface du globe et sur les changements qu'elles ont produits dans le règne animal* (Paris: C. Bourgois, 1985), p. 96.
2. See Wellek, *Kant in England.*
3. See Appell, *The Cuvier–Geoffrey Debate.*
4. Whewell, quoted in J. M. Lynch, *Vestiges and the Debate before Darwin* (London: Thoemmes Press, 2000), p. 89.
5. See N. Rupke, *Richard Owen: Biology without Darwin* (Chicago, IL: University of Chicago Press, 2009). Darwin could not resist the Germanophilia that spread in his country during the 1930s. Returned from his voyage on the *Beagle*, he tried to learn German with his brother Erasmus, who had previously studied in Germany (Owen, *Hunterian Lectures*).

6. P. Rehbock, *The Philosophical Naturalists: Themes in Early Nineteenth-Century British Biology* (Madison, WI: University of Wisconsin Press, 1983).
7. Ibid., p. 53.
8. Rupke, *Richard Owen*, p. 114.
9. Rehbock, *The Philosophical Naturalists*, p. 59.
10. K. E. von Baer, 'Fragments Related to Philosophical Zoology: Selections from the Works of K. E. von Baer, trans. T. H. Huxley', in A. Henfrey and T. Huxley (eds), *Scientific Memoirs, Selected from the Transactions of Foreign Academies of Science* (London: Taylor & Francis, 1853), pp. 176–238, on p. 177.
11. Russell, *The Interpretation of Development*, p. 37.
12. Lenoir, *The Strategy of Life*, p. 114.
13. Von Baer, quoted in Oppenheimer, *Autobiography*, p. 181.
14. Von Baer, quoted in Russell, *The Interpretation of Development*, p. 37.
15. See E. S. Russell, *Form and Function: A Contribution to the History of Morphology* (London: J. Murray, 1916), p. 129.
16. Ibid., p. 49.
17. Cassirer, *The Problem of Knowledge*, pp. 129–30. Lenoir too had no doubts: 'Cuvier was responsible for deepening and extending the teleomechanist program of vital materialism' (Lenoir, *The Strategy of Life*, p. 54).
18. See D. Outram, 'Uncertain Legislator: Georges Cuvier's Laws of Nature in their Intellectual Context', *Journal of the History of Biology*, 19:3 (1986), pp. 323–68; and P. Taquet, *Georges Cuvier, Naissance d'un Génie* (Paris: Odile Jacob, 2006).
19. Cuvier, quoted in Outram, 'Uncertain Legislator', p. 34.
20. Lenoir, *The Strategy of Life*, p. 55.
21. See also D. Ospovat, 'The Influence of Karl Ernst von Baer's Embryology, 1828–1859: A Reappraisal in Light of Richard Owen's and William B. Carpenter's Paleontological Application of von Baer's Law', *Journal of the History of Biology*, 9:1 (1976), pp. 1–28.
22. Desmond mentions a very revealing letter that Owen sent to Whewell in 1840: 'how much we – i.e. the Cuvieriean cultivators of comparative anatomy – are, and always must be, indebted to you for the clear statement of the scientific character of teleological thinking'; A. Desmond, *The Politics of Evolution: Morphology, Medicine, and Reform in Radical London* (Chicago, IL: University of Chicago Press, 1989), p. 85. And yet, as he further reminds us: 'it is clear that German nature-philosophers (such as Oken and Carus) and "vital materialists" (such as Meckel and the transformationist Tiedemann) had a much broader influence in Britain on the philosophical anatomists and their allies (many – Knox, Shapey, Grant, Turner, Marshall Hall, Martin Barry – had attended courses in Germany)'; ibid.
23. About Coleridge, as Ritterbush describes: 'In a number of lectures and essays, of which the most important was "On the Definitions of Life" (1830), Coleridge criticized empirical physiologists as pursuing too narrow an approach to the problems of life. They studied organisms as though they were simple physico-chemical assemblages, a piece at a time, while the most important phenomena of life resided at the level of the organism as a whole'; P. C. Ritterbush *The Art of Organic Form* (Washington, DC: Smithsonian Institution Press, 1968), p. 16. Yet, like Green, Coleridge accepted all the tenets of the post-Kantian bio-philosophy: 'the origin of the whole precedes the differentiation of the parts. The whole is primary; the parts are derived ... Second, the form manifests the process of growth by which it arose ... The organic form proclaims itself as the end result of a progressive sequence of development. Third, as it grows the plant assimilates diverse

elements into its own substance ... Fourth, the achieved form of the plant is directed from within ... The external aspect of living things is determined by internal processes, not, as in a human artefact, from without. Fifth, the parts of the living whole are interdependent' (see ibid., pp. 20–4).

24. P. R. Sloan, 'Kant and British Bioscience', in P. Huneman (ed.), *Understanding Purpose: Kant and the Philosophy of Biology* (Rochester, NY: University of Rochester Press, 2007), pp. 149–70, on p. 155.
25. Owen deemed Whewell's philosophy the nineteenth-century *Novum Organon* of Bacon; Rupke, *Richard Owen*.
26. See ibid.
27. Owen, *Hunterian Lectures*, p. 214.
28. See M. Ruse, *The Darwinian Revolution: Science Red in Tooth and Claw* (Chicago, IL: University of Chicago Press, 1999).
29. See Cassirer, *The Problem of Knowledge*, p. 204.
30. Owen, *Hunterian Lectures*, p. 72.
31. Ospovat, 'The Influence of Karl Ernst von Baer's Embryology', p. 24.
32. See M. Goodman, *Suffer and Survive* (London: Simon & Schuster, 2007).
33. J. Pagel, *Biographical Encyclopedia of Outstanding Physicians of the Nineteenth Century*, Vol. 1550–1 (Vienna: Verlag Urban and Schwarzenberg, 1901).
34. Anon., 'Obituary, William Thierry Preyer (1841–1897)', *Nature*, 56:1448 (1897), p. 296.
35. W. T. Preyer, *Eléments de physiologie générale* (Paris: Alcan, 1884), p. 4.
36. Ibid., pp. 4–5.
37. Ibid., p. 97.
38. E. Schwarz-Weig, 'Science in a Constant Flow: Life and Work of Eduard Strasburger (1844–1912)', *German Botanical Society*, 2002, at http://www.deutsche-botanische-gesellschaft.de [accessed 13 August 2013].
39. E. Strasburger, *A Text-Book of Botany* (London: Macmillan, 1898), p. 160.
40. Ibid., p. 161.
41. Haldane first went to Germany in 1879, to Jena, where he attended Haeckel's lectures, and then to Fribourg, where he aimed to improve his 'microscope skills', attending Weismann's lectures (Goodman, *Suffer and Survive*, p. 106). By that time he had already visited the country several times for a variety of reasons.
42. See Sturdy, 'Biology as Social Theory'. See also P. Robbins, *The British Hegelians 1875–1925* (London: Garland, 1982).
43. See Chapter 3 in this volume.
44. I should mention that in 1888 Geddes held the chair of botany at the University College of Dundee. He was, as we shall see, highly influential on E. S. Russell's biological thought.
45. See J. S. McDowall, 'Sir Michael Foster', *Post-Graduate Medical Journal*, 12 (1936), pp. 78–9; and G. L. Geison, *Michael Foster and the Cambridge School of Physiology: The Scientific Enterprise in Late Victorian Society* (Princeton, NJ: Princeton University Press, 1978).
46. M. Foster and F. M. Balfour, *The Elements of Embryology* (London: Macmillan, 1874), p. 6.
47. See B. K. Hall, 'Francis Maitland Balfour (1851–1882): A Founder of Evolutionary Embryology', *Journal of Experimental Zoology*, 299 (2003), pp. 3–8; and M. Richmond, 'A Lab of One's Own: The Balfour Biological Laboratory for Women at Cambridge University, 1884–1914', *ISIS*, 88:3 (1997), pp. 422–55.

48. H. Blackman, 'A Spiritual Leader? Cambridge Zoology, Mountaineering and the Death of F. M. Balfour', *Studies in History and Philosophy of Science Part C: Studies in History and Philosophy of Biological and Biomedical Sciences*, 35:1 (2003), pp. 93–117.
49. R. D. Thompson, *D'Arcy Wentworth Thompson, The Scholar-Naturalist, 1860–1948* (Oxford: Oxford University Press, 1958), p. 52.
50. Ibid., pp. 70–1.
51. A. Richmond, "Sedgwick, Adam (1854–1913)', *Oxford Dictionary of National Biography* (Oxford: Oxford University Press, 2004), p. 365
52. Ibid., p. 366.
53. A. Sedgwick, *Student's Text-Book of Zoology* (London: Macmillan, 1898), p. 11.
54. Ibid., p. 10.
55. Ibid.
56. See B. Lightman, *The Dictionary of Nineteenth-Century British Scientists* (London: Thoemmes Continuum, 2004), p. 1991.
57. See Bowler, *Reconciling Science and Religion*.
58. J. A. Thomson, *The Study of Animal Life* (London: J. Murray, 1901), p. 141.
59. See J. A. Thomson, *Systems of Animate Nature: The Gifford Lectures Delivered in the University of St. Andrew in the Years 1915 and 1916*, 2 vols (London: Williams and Norgate, 1920).
60. Books such as *Life: Outlines of General Biology*; see P. Kitchen, *A Most Unsettling Person: An Introduction to the Ideas and Life of Patrick Geddes* (London: Gollancz, 1975).
61. See J. A. Thomson and P. Geddes, 'A Biological Approach', in J. E. Hand (ed.), *Ideals and Science and Faith* (London: George Allen, 1904).
62. M. Graham, 'E. S. Russell 1887–1954', *ICES Journal of Marine Science*, 20:2 (1954), pp. 135–9, on p. 138.
63. See Chapter 4 in this volume.
64. See J. R. Gregg and F. T. C. Harris, *Form and Strategy in Science: Studies Dedicated to Joseph Henry Woodger on the Occasion of his Seventieth Birthday* (Dordrecht: D. Reidel, 1964).
65. See D. M. S. Watson, 'James Peter Hill, 1873–1954', *Biographical Memoirs of Fellows of the Royal Society*, 1 (1955), pp. 101–17.
66. See B. D. J., 'Dr. John Beard', *Nature*, 114:904 (1924), p. 904.
67. See D. R. Cohen, 'Living Precisely in the Fin-de-Siècle Vienna', *Journal of the History of Biology*, 39:3 (2006), pp. 493–532.
68. See A. Koestler, *The Case of the Midwife Toad* (London: Hutchinson, 1971).
69. D'Arcy Thompson, Paul Weiss, Karl Frisch, Gregory Bateson and Eugen Steinach all worked there. The importance of this centre has been emphasized by Cohen in 'Living Precisely'.
70. Cohen, 'Living Precisely', p. 496.
71. See Gregg and Harris, *Form and Strategy*, p. 3.
72. B. Kuklick, *A History of Philosophy in America: 1720–2000* (Oxford: Oxford University Press, 2001), p. 64.
73. See ibid. In particular, James was inspired by Royce's *Religious Aspects of Philosophy* (1913).
74. See B. Kuklick, *Josiah Royce: An Intellectual Biography* (Indianapolis, IN: Hackett Publishing Company, 1985).
75. J. Dewey, *The Influence of Darwin on Philosophy and Other Essays* (New York: Henry Holt, 1910), p. 19.

76. W. Coleman, *Biology in the Nineteenth Century: Problems of Form, Function and Transformation* (Cambridge: Cambridge University Press, 1971).
77. See G. E. Allen, *Life Science in the Twentieth Century* (New York: Wiley, 1975); and J. Maienschein, 'The Origins of Entwicklungsmechanik', in S. F. Gilbert (ed.), *A Conceptual History of Modern Embryology* (New York: Plenum Press, 1991), pp. 43–59.
78. See K. Sander, 'Wilhelm Roux and the Rest: Developmental Theories 1885–1895', *Roux's Archives of Developmental Biology*, 200 (1991), pp. 293–9.
79. An institution directed by a student of Haeckel, Anton Dohrn. See I. Müller, 'The Impact of the Zoological Station in Naples on Developmental Physiology', *International Journal of Developmental Biology*, 40 (1996), pp. 103–11.
80. Nyhart, *Biology Takes Form*, p. 96.
81. L. K. Nyhart and S. Lidgard, 'Individuals at the Center of Biology: Rudolf Leuckart's *Polymorphismus* and the Ongoing Narrative of Parts and Wholes. With an Annotated Translation', *Journal of the History of Biology*, 44:3 (2011), pp. 373–443, on p. 378.
82. See Anon., 'Rudolf Leuckart', *Nature*, 57:1484 (1898), p. 542; and R. Blanchard, 'Notices biographiques: Rodolphe Leuckart', *Archives de parasitologie*, 1 (1898), pp. 185–90.
83. The Leuckart international community comprised, apart from Americans, students from England, France, Italy, Sweden, Russia, Switzerland and Japan. See H. Schadewaldt, 'Leuckart, Karl Georg Friedrich Rudolf', *Encyclopedia.com*, at http://www.encyclopedia.com/topic/Karl_Georg_Friedrich_Rudolf_Leuckart.aspx [accessed 13 August 2013].
84. See Nyhart and Lidgard, 'Individuals'; as the letters Leuckart exchanged with von Baer and that Lenoir quotes attest.
85. For a detailed discussion of functional morphology in Germany, see Lenoir, *The Strategy of Life*, pp. 157–94.
86. Ibid., p. 217.
87. Nyhart, *Biology Takes Form*, p. 340.
88. E. S. Morse, *Biographical Memoir of Charles Otis Whitman, 1842–1910* (Washington, DC: National Academy of Sciences, 1912).
89. P. M. Winsor, *Reading the Shape of Nature: Comparative Zoology at the Agassiz Museum* (Chicago, IL: University of Chicago Press, 1991), p. 174.
90. See M. Beckner, *The Biological Way of Thought* (Berkeley, CA: University of California Press, 1959).
91. H. Bergson, *Creative Evolution* (London: Macmillan, 1922), p. 268.
92. See R. Bud, *The Uses of Life: A History of Biotechnology* (Cambridge: Cambridge University Press, 1993), and Bowler, *Reconciling Science and Religion*.
93. A. N. Whitehead, *Science and the Modern World* (New York: Free Press, 1997), p. 135.
94. Ibid.
95. See J. S. Huxley, *Evolution: The Modern Synthesis* (London: Allen & Unwin, 1954).
96. See E. Mayr, *Toward a New Philosophy of Biology: Observations of an Evolutionist* (Cambridge, MA: Harvard University Press, 1988).

## 3 The British Version: J. S. Haldane, D'Arcy Thompson and the Organism as a Whole

1. As Goodman reports, Haldane was highly enthusiastic about his German training (Goodman, *Suffer and Survive*, p. 67).

2. As Goodman pointed out: 'A century had passed, but Haldane now found himself in both the intellectual and physical space of Goethe' (ibid, p. 65).
3. Robert Haldane was mainly a philosopher. He also trained in Germany, and he became the first translator of, and expert in, Schopenhauer's philosophy in Britain.
4. J. S. Haldane and R. S. Haldane, 'The Relation of Philosophy to Science', in A. Seth and R. S. Haldane (eds), *Essays in Philosophical Criticism* (New York: Burt Franklin, 1883), pp. 41–66, on pp. 45–55.
5. In 1893 Herdman would welcome into his Liverpool department a young American biologist, William Emerson Ritter, during his first trip in Europe. See B. R. N. Rudmose, 'Sir William Abbott Herdman (1858–1924)', *Oxford Dictionary of National Biography* (Oxford: Oxford University Press, 2004); see also Chapter 5 in this volume.
6. Thompson, *D. W. Thompson*, p. 25.
7. Haldane built a device through which external air passed and deposited microscopic germs in a jelly substance that worked as a substrate for the growth of microbial colonies; so that, once back in the lab, he could study and see the level of pollution and air contamination breathed by dwellers.
8. Sherrington was awarded the Nobel Prize in 1932 for his work in physiology; see F. C. Rose, *Twentieth-Century Neurology: The British Contribution* (London: Imperial College Press, 2002).
9. As his daughter Ruth Thompson also recalls (*D. W. Thompson*, pp. 194–5), there is a very revealing letter about Bergson sent by Thompson to P. Geddes: 'I envy your recent trip, and especially your talk with Bergson and others. I sent, by the way, a copy of my B.A. Address to Bergson, and got a very kindly note from him in reply. As matter of fact, the good man is so well boomed at present, at least in England, that there is not the least fear of his light being hid under a bushel ... I should infinitely rather have him, if it be found possible, as a Gifford Lecturer, than the various Totemists and myth-mongers whom we are so well accustomed to' (Thompson to Geddes, 25 October 1911, letter ms16400, Thompson's papers, University of St Andrews Archives (hereafter DTP-SAA)). Geddes was another figure belonging to the close group of Thompson's friends – a group that included Whitehead himself, with whom Thompson was in correspondence. About the influence of Bergson on biological sciences, Geddes made clear (in a document dated 1914 and addressed to the Secretary of the University of St Andrews) that the French professor's fame was a sign that biological and evolutionary studies – after a period of crisis – were thriving again; see Geddes, 22 January 1914, letter ms16428, DTP-SAA.
10. At a very early age, D'Arcy Thompson translated into English Hermann Müller's *Fertilization of Flowers*, with an introduction by Charles Darwin. See H. Müller, *The Fertilization of Flowers*, ed. and trans. D. W. Thompson, with a preface by C. Darwin (London: Macmillan, 1883).
11. In this piece, Thompson was answering an article that Haldane had published a few months earlier in *Mind*; see J. S. Haldane, 'Life and Mechanism', *Mind*, 9:33 (1884), pp. 27–47.
12. D. W. Thompson, 'The Regeneration of Lost Parts in Animals', *Mind*, 9:35 (1884), p. 419.
13. Ibid.
14. J. S. Haldane, D. W. Thompson, P. C. Mitchell and L. T. Hobhouse, 'Symposium: Are Physical, Biological and Psychological Categories Irreducible?', *Proceedings of the Aristotelian Society: Supplementary Volumes*, new ser., 18 (1918), pp. 419–78, on p. 422.
15. Ibid., p. 30.

16. Ibid., p. 48.
17. As he would say in his *Growth and Form:* 'even the warm-blooded animal is not in reality a heat-engine; working as it does at almost constant temperatures its output of energy is bound, by the principle of Carnot, to be small. Nor is it an electrostatic machine, nor yet an electrodynamic one. It is a mechanism in which chemical energy turns into surface-energy, and working hand in hand, the two are transformed into mechanical energy, by steps which are for the most part unknown'; D. W. Thompson, *On Growth and Form* (Cambridge: Cambridge University Press, 1942), p. 464.
18. Haldane et al., 'Symposium', p. 46.
19. Ibid., p. 73.
20. D. W. Thompson, 'Review of J. S. Haldane's The New Physiology', *Mind*, 27 (1919), pp. 359–61, on p. 361.
21. Ibid., p. 362.
22. See the British Association for the Advancement of Science, Report of the Ninety-Seventh Meeting, South Africa, 22 July–3 August 1929, Office of the British Association, Burlington House, London, 1930. As Thompson's daughter, Ruth, reports: 'This meeting was packed to the doors and many people were turned away' (Thompson, *D. W. Thompson*, p. 194).
23. As Goodman describes the relationship between Smuts and Haldane: 'Haldane met many times with Jan Smuts and his wife Isie, getting along famously. When Smuts visited Oxford later that year for a series of sell-out lectures at the Sheldonian, Haldane gave him guidance as to how to handle English audiences. Such was Smuts's charm that he managed to bring Kathleen into his and Haldane's holistic way of thinking' (Goodman, *Suffer and Survive* p. 365). Yet Thompson's daughter, Ruth, records that her father first met Smuts in South Africa. She adds that Smuts was a scholar: 'he had always greatly admired' (Thompson, *D. W. Thompson*, p. 194).
24. See Goodman, *Suffer and Survive*, p. 366. A few years later Haldane recalled: 'In his Presidential Address to the British Association (1931) he gave (Smuts) a summary of his reasoning, and particularly of its relations to the interpretation of evolution, which he regards as the unfolding of holistic reality in what had commonly been regarded as purely mechanic universe'; J. S. Haldane, *The Philosophy of a Biologist* (Oxford: Clarendon Press, 1935), pp. 40–1.
25. See Haldane, *The Philosophical Basis of Biology*, p. 133.
26. Haldane, *The Philosophy of a Biologist*, p. 5.
27. See Sturdy, 'Biology as Social Theory'.
28. E. P. Robins, *Some Problems of Lotze's Theory of Knowledge* (New York : Macmillan, 1900), p. 11.
29. Haldane, 'Life and Mechanism', pp. 27–47.
30. Ibid., p. 35.
31. Haldane reports an experiment in which a frog's spinal cord had been severed from the brain. When an irritating substance was applied to the skin of the frog's limb, the limb tried to get rid of such a substance. If the leg was blocked, the other leg tried to get rid of the irritating substance on the first leg. This demonstrated, Haldane argued, that there was not a direct relation between the stimulus and the response happening between brain, spinal cord and skin receptors, and that a mechanist viewpoint as applied to that problem was too simplistic in approaching a far more complex phenomenon.
32. As Haldane wrote: 'Müller had himself taken a prominent part in disproving by his microscopical observations on glands the mechanistic theories previously prevailing

with regard to secretion, and his numerous observations on growth and development had led him in the same direction'; J. S. Haldane, *Mechanism, Life and Personality: An Examination of the Mechanistic Theory of Life and Mind* (London: Murray, 1913), p. 42. Yet as he stated in the preface of the first edition of his large 1935 monograph *The Philosophy of a Biologist* in relation to the work of Müller and other old physiologists: 'My treatment of the subject may possibly be looked on askance in some quarters as reactionary; for I have been largely influenced by the ideas and work of older physiologists. If, however, I have gone backwards, it is only to pick up clues which had been temporarily lost; and all these clues lead forward – forward to a new physiology which embodies what was really implicit in the old' (Haldane, *The Philosophy of a Biologist*, p. x).

33. Haldane went on for a few pages in criticizing Weismann: 'The mechanistic theory of heredity is not merely unproven, it is impossible, it involves such absurdities that no intelligent person who has thoroughly realized its meaning and implications can continue to hold' (Haldane, *The Philosophy of a Biologist*, pp. 54–8).
34. Ibid., p. 124.
35. J. S. Haldane, *Materialism* (London: Hodder and Stoughton, 1932), p. 38.
36. Consisting of a cage with a canary or a mouse inside, it was a device used to warn miners about the presence of carbon monoxide or other poisonous gases.
37. In 1905 Haldane was appointed by the Royal Navy, and in particular by the Deep Sea Diving Committee, in order to perform research on deep diving technical problems. In fact, the main aim was to understand how to avoid the so-called 'caisson disease', or, as we would say now, decompression disease. In 1907 he published his decompression tables, which were widely used until the late 1950s.
38. See G. E. Allen, 'J. S. Haldane: The Development of the Idea of Control Mechanisms in Respiration', *Journal of the History of Medicine*, 22 (1967), pp. 392–412; see also G. J. Goodfield, *The Growth of Scientific Physiology* (London: Hutchinson, 1960).
39. J. S. Haldane, *The New Physiology and Other Addresses* (London: C. Griffin, 1919), p. 40.
40. Ibid.
41. See ibid., p. 25.
42. J. S. Haldane, *Organism and Environment as Illustrated by the Physiology of Breathing* (New Haven, CT: Yale University Press, 1917), p. 59.
43. Ibid., p. 97.
44. Haldane, *Materialism*, p. 36.
45. Haldane, *The New Physiology*, p. 86.
46. Ibid., p. 87.
47. Ibid., p. 89.
48. The essay entitled 'The Institutes of Medicine and Surgery' was based on a paper he delivered at the Middlesex Medical Society and published in his 1932 book *Materialism*.
49. Haldane, *Materialism*, p. 42.
50. Ibid., p. 46.
51. Haldane, *The Philosophical Basis of Biology*, pp. 27–8.
52. Haldane, *Mechanism, Life and Personality*, p. 60.
53. Haldane, *Materialism*, p. 62.
54. Ibid., p. 122.
55. Ibid., p. 124.
56. In this book, Haldane developed one of his most extensive critiques and discussions of Kant's philosophy. He started with quite a detailed discussion of Kant's *Critique of Pure Reason*, and then discussed post-Kantian philosophies (Hegel, Fichte, Schopenhauer,

etc.). Finally, he introduced a long discussion about Kant's *Critique of Judgment*. In support of his holistic stance, Haldane cited Whitehead's philosophy of physics and J. C. Smuts's *Holism and Evolution* (New York: Macmillan, 1926).

57. Haldane, *The Philosophy of a Biologist*, p. 74.
58. In the Silliman lectures delivered at Yale University in 1915, Haldane clearly made the connection between his stance and Delage's organicism as expressed in his book *L'Hérédité et les grandes problèmes de la biologie générale*, 2nd edn (Paris: Reinwald, 1903). See Haldane, *Organism and Environment*, p. 2.
59. Haldane, quoted in Sturdy, 'Biology as Social Theory', p. 322.
60. See M. Wolff, 'Hegel's Organicist Theory of the State: On the Concept and Method of Hegel's Science of the State', in R. Pippin (ed.), *Hegel on Ethics and Politics* (Cambridge: Cambridge University Press, 2004), pp. 291–322. Also see Beiser, *Hegel.*
61. Haldane, *Materialism*, p. 195.
62. Ibid., p. 197.
63. Ibid., p. 198.
64. See M. Carter, *T. H. Green and the Development of Ethical Socialism* (Exeter: Imprint Academic, 2003).
65. As he concluded his talk: 'My own work in pure science can be summed up in the conclusion that the mechanical conceptions of physical science break down irretrievably when we endeavour to apply them to life and conscious behaviour, both of which are just a part of what we call Nature; and I have just tried to point out that mere economic conception of industrial life breaks down equally hopelessly when we try to apply it to such an industrial undertaking as a British coal-mine' (Haldane, *Materialism*, p. 204).
66. Sturdy, 'Biology as Social Theory', p. 334.
67. S. J. Gould, *The Structure of Evolutionary Theory* (Cambridge, MA: Harvard University Press, 2002), p. 1204. On D'Arcy Thompson's supposed isolation, see also S. J. Gould, 'D'Arcy Thompson and the Science of Form', *New Literary Hist*ory, 2:2 (1971), pp. 229–58.
68. See J. Whitfield, *In the Beat of a Heart: Life, Energy and the Unity of Nature* (Washington, DC: Joseph Henry Press, 2006); A. Wallace, 'D'Arcy Thompson and the Theory of Transformation', *Nature Review Genetics*, 7:5 (2006), pp. 401–6; D. W. Thompson, *On Growth and Form*, ed. J. T. Bonner, rev. and abridged edn (Cambridge: Cambridge University Press, 1966); and M. Ridley, 'Embryology and Classical Zoology in Great Britain', in T. J. Horder, J. A. Witkowski and C. C. Wylie (eds), *A History of Embryology* (Cambridge: Cambridge University Press, 1986), pp. 35–68.
69. See J. Pasteels, 'Notice sur Albert Dalcq', *Annuaire de l'Academie Royale de Belgique*, 129 (1974), pp. 155–66; J. G. Mulnard, "The Brussels School of Embryology', *International Journal of Developmental Biology*, 36 (1992), pp. 17–24; and D. Thieffry, 'Rationalizing Early Embryogenesis in the 1930s: Albert Dalcq on Gradients and Fields', *Journal of the History of Biology*, 34 (2001), pp. 149–81.
70. Dalcq to the Secretary of the Royal Society of Edinburgh, 6 June 1939, letter ms25711, hereafter DTP-SAA.
71. See, in particular, Dalcq to Thompson, letters ms2885 and ms25712, DTP-SAA.
72. See O. Stork, 'Berthold Hatschek: A Landmark in the History of the Morphology of the Austrian Academy of Sciences', *Almanac for the year*, 99 (1946), p. 284.
73. Przibram to Thompson, 1927, letter ms25830, DTP-SAA.
74. Particularly in the second edition of *On Growth and Form*, Thompson referred to several of Przibram's works. As Drack, Apfalter and Pouvreau write, Przibram in Vienna

'committedly discusses the possibility of applying mathematics to biology in general and to morphogenesis in particular. He also considers laws of higher order in cases where the detailed underlying single processes are not known. "Anyway, the influence of the 'whole' on the 'part' turns out to be accessible to mathematical processing, grounded on our ideas from exact natural history [*Naturlehre*]". Thus, he proposes a "mathematical morphology" by which the parameters of the developing forms (e.g., proportions) should be described, also in relation to physical measures such as surface tension. The investigation should reveal "constants" that explain similar forms in various species ... the fundamental idea here is to unite morphology and physiology, an improvement compared to D'Arcy W. Thompson's approach of mathematical morphology'; M. Drack, W. Apfalter and D. Pouvreau, 'On the Making of a System Theory of Life, Paul A. Weiss and Ludwig von Bertalanffy's Conceptual Connections', *Quarterly Review of Biology*, 82:4 (2007), pp. 349–73, on p. 354.

75. Cohen, 'Living Precisely', p. 494.
76. See H. Przibram, *Aufbau mathematischer Biologie*, Abhandlungen zur theoretischen Biologie, Vol. 18 (Berlin: Gebrüder Borntraeger, 1923).
77. Przibram to Crew, 14 November 1928, letter ms17234, DTP-SAA.
78. See Thompson, *On Growth and Form*, p. 782.
79. E. N. Andreade, 'Obituary of E. Hatschek', *Nature*, 154:3897 (1944), pp. 46–7.
80. From 1910 Hatschek retired from any professional occupation.
81. Thompson, to Hatschek, 31 July 16, letter ms26323, DTP-SAA.
82. Hatschek to Thompson, 12 August 1918, letter ms26362, DTP-SAA.
83. See, for example, letter ms26327, Hatschek to Thompson, 16 August 1918, DTP-SAA.
84. Thompson to Hatschek, 24 October 1918, letter ms26329, DTP-SAA.
85. In 1918 the Proceeding of the Royal Society of London published a paper that Hatschek had written with Thompson's suggestions and help; see E. Hatschek, 'A Study of the Forms Assumed by Drops and Vortices of a Gelatinising Liquid in Various Coagulating Solutions', *Proceedings of the Royal Society of London. Series A, Containing Papers of a Mathematical and Physical Character (1905–1934)*, 95:669 (1918), pp. 303–16.
86. Conklin to Thompson, 23 December 1945, letter ms24976, DTP-SAA.
87. Conklin to Thompson, 6 August 1942, letter ms42364, DTP-SAA.
88. B. H. Willier, 'Frank Rattray Lillie, 1870–1947', in *Biographical Memoirs: National Academy of Sciences* (Washington, DC: National Academies Press, 1957), at http://www.nasonline.org/publications/biographical-memoirs/memoir-pdfs/lillie-frank-r.pdf [accessed 13 August 2013]. Also see Chapter 5 in this volume.
89. Thompson to Lillie, 8 June 1942, Box IIA, Folder 84, Lillie Correspondence, Marine Biological Library Historical Archives, Woods Hole Oceanographic Institution, MA, USA (hereafter MBL Archive, Woods Hole).
90. Lillie to Thompson, 15 July 1942, Box IIA, Folder 84, Lillie Correspondence, MBL Archive, Woods Hole.
91. Lillie read the new edition of *On Growth and Form* when he received it in Woods Hole; indeed, on 12 October 1942 he wrote: 'I immediately dipped into it, and wish to congratulate you on the completion of such a monumental task of scholarship. Who else but you could span the gap from Aristotle to the present with apposite citations spread along the way!'; Lillie to Thompson, 12 October 1942, Box IIA, Folder 84, Lillie Correspondence, MBL Archive, Woods Hole).
92. See K. R. Dronamraju, *If I Am to be Remembered: The Life and Work of Julian Huxley with Selected Correspondence* (Singapore: World Scientific, 1993).

93. See J. Gayon, 'History of the Concept of Allometry', *American Zoologist*, 40:5 (2000), pp. 748–58.
94. J. S. Huxley and G. Teissier, 'Terminology of Relative Growth', *Nature*, 137 (1936), pp. 780–1.
95. Thompson to Huxley, 5 March 1925, letter ms28563, DTP-SAA.
96. Thompson to Huxley, 6 March 1925, letter ms28564, DTP-SAA.
97. Huxley to Thompson, 29 January 1932, letter ms10065, DTP-SAA.
98. J. S. Huxley, *Problems of Relative Growth* (London: Methuen, 1932), p. xi.
99. See, for instance, Huxley to Thompson, 31 March 1931, letter ms10062, DTP-SAA.
100. Huxley to Thompson, 11 November 1941, letter ms25770, DTP-SAA.
101. Thompson, *On Growth and Form*, p. 1.
102. Gould, *The Structure*, p. 1204.
103. N. B. Ninham and P. Lo Nostro, *Molecular Forces and Self-Assembly* (Cambridge: Cambridge University Press, 2010).
104. See Whitfield, *In the Beat of a Heart*; W. Le Gros Clark and and P. Medawar, *Essays on Growth and Form: Presented to D'Arcy Wentworth Thompson* (Oxford: Clarendon Press, 1945); and L. Kay, *The Molecular Vision of Life: Caltech, the Rockefeller Foundation, and the Rise of the New Biology* (Oxford: Oxford University Press, 1993).
105. In Germany and Austria, for instance, influential scholars such as Adolf Meyer Abich (1893–1971) and von Bertalannfy (1901–72) highlighted the use of mathematics for solving or describing biological phenomena. See K. S. Amidon, 'Adolf Meyer-Abich, Holism, and the Negotiation of Theoretical Biology', *Biological Theory: Integrating Development, Cognition and Evolution*, 3:4 (2008) pp. 357–70; see also L. von Bertalanffy, *Modern Theories of Development: An Introduction to Theoretical Biology* (Oxford: Oxford University Press, 1933); and A. Meyer-Abich, *Ideen und Ideale der biologischen Erkenntnis: Beitrage zur Theorie und Geschichte der biologischen Ideologien* (Leipzig: J. A. Barth, 1934).
106. G. K. Kimball, 'D'Arcy Thompson: His Conception of the Living Body', *Philosophy of Science*, 20:2 (1953), pp. 139–48, on p. 143.
107. Ibid., p. 141.
108. Medawar, quoted in Thompson, *D. W. Thompson*, p. 227.
109. He admired Aristotle: he had translated *Historia Animalium*, a *Glossary of Greek Fishes* and *A Glossary of Greek Birds.*
110. Thompson, *D. W. Thompson*, p. 222.
111. See Bonner's introduction to the abridged edition of Thompson's *Growth and Form.*
112. Thompson, *On Growth and Form*, p. 270.
113. Ibid., p. 849.
114. Ibid., p. 960.
115. Ibid., p. 1117.
116. Goethe, quoted in Cassirer, *The Problem of Knowledge*, p. 204.
117. Thompson, *On Growth and Form*, p. 7.
118. Ibid., p. 10.
119. The term 'mechanism' for Thompson did not refer to a mechanical engine such as a steam-engine or an automaton, but to a biological mechanism. In fact, with this notion, he meant an entity that did not defy physical and chemical laws.
120. Thompson, *On Growth and Form*, p. 14.
121. 'When a living organism is broken down into its elements, you cannot put these elements together and expect that the organism would return to life again' (ibid., p. 344 (my translation)).

122. Ibid., p. 274.
123. Ibid., p. 282.
124. Ibid., p. 284.
125. De Bary's famous remark towered at the end of the chapter (ibid., p. 345). Whitman's oft-quoted passage concluded: 'the fact that physiological unity is not broken by cell-boundaries is confirmed in so many ways that it must be accepted as one of the fundamental truths in biology' (Whitman, quoted in ibid.).
126. Ibid.
127. Bergson had defined living beings as 'centres of action'; H. Bergson, *L'Évolution Créatrice* (Paris: Les Presses universitaires de France, 1907), p. 262.
128. Goodsir, as Thompson reported, had already used the metaphor of cell as centre of growth and force a century before. See Thompson, *On Growth and Form*, p. 286.
129. Ibid., p. 333.
130. Ibid., p. 342.
131. See J. Sapp, *Beyond the Gene: Cytoplasmic Inheritance and the Struggle for Authority in Genetics* (Oxford: Oxford University Press, 1987); and Thompson, *On Growth and Form*, p. 464.
132. As Bonner argued: 'Heredity and activity of geneses in development are wholly missing in the book except for few passing references which seems to imply that they do not fit into his scheme of things' (Thompson, *Growth and Form*, ed. Bonner, p. xviii). However, in my opinion, this is only partially true. Indeed, to Thompson, heredity was not synonym of genetics.
133. Thompson, *On Growth and Form*, p. 289.
134. Ibid., p. 288.
135. Ibid.
136. In the archives of St Andrews are conserved many of Thompson's handwritten and typewritten notes and papers about Mendelian genetics. Although Thompson never published on the topic, he was deeply interested. He gathered quotations taken from diverse geneticists and studied Punnett's works about the relation between disease and Mendelian heredity. He was profoundly interested in the application of the principles of Mendelian genetics in animal breeding. In particular, he studied A. D. Darbishire's *Recent Advances in Animal Breeding and their Bearing on our Knowledge of Heredity* (1907). Finally, he did not underestimate, during the very last years of his life, what was happening in Soviet Russia with Ivan Minchurin's approach to heredity.
137. Thompson, quoted in Whitfield, J., *In the Beat of a Heart*, p. 20.
138. Thompson, *On Growth and Form*, p. 1019.
139. 'The cause lying behind the existence of every part of the living body is included into its whole' (my translation); J. Müller, *Handbuch der Physiologie des Menschen: für Vorlesungen*, Vol. 1 (Coblenz: Verlag von J. Holscher, 1837), p. 18; Thompson, *On Growth and Form*, p. 1020.
140. Thompson, *On Growth and Form*, p. 1018.
141. Ibid.
142. Ibid., p. 1022.
143. Ibid., p. 1025.
144. Ibid., p. 1037.
145. Ibid., p. 1094.
146. Ibid., p. 1095.

147. Richards, in reconstructing the intellectual context behind Haeckel's famous work on *Radiolaria*, recalls that 'The idea that descent relationship might operate according to various mathematical deformations of the basic sphere was quite in the older Goethean tradition of morphology, comparable to Carus's derivation of the form of the vertebra from geometrical arrangements of the basic sphere' (Richards, *The Tragic Sense of Life*, p. 94).
148. Kant, *Kritik of Judgment*, p. 280.
149. Medawar, quoted in Thompson, *D. W. Thompson*, p. 221.
150. As Foster clearly demonstrated in writing to Thompson in 1894: 'If the form is constant in a group – it does not matter how the form is brought about' (quoted in Thompson, *D. W. Thompson*, p. 90).

## 4 The New Generation: A Failed Organismal Revolution

1. The Second London Congress for the History of Science also took placed in 1931. As we have seen in the previous chapter, Haldane participated in a session chaired by Ritter and attended by Russell and Woodger, both of whom contributed.
2. Haldane, *The Philosophical Basis of Biology*, p. 148.
3. Ibid., pp. 149–50.
4. Ibid., pp. 152–3.
5. E. S. Russell, *The Directiveness of Organic Activities* (Cambridge: Cambridge University Press, 1945), p. 176.
6. Letter from von Bertalanffy to Ritter, 22 October 1931, Ritter Papers, 71/3 c, Box 6, Folder: Bertalanffy 1901, Bancroft Library Archive, University of Berkeley, California (hereafter Bancroft Archive). The translation of the letter is my own from the German.
7. Quoted in von Bertalanffy, *Modern Theories*, p. 4.
8. See Graham, 'E. S. Russell'.
9. Ibid., p. 138.
10. See the archive of PhD theses at the University of Glasgow, at http://eleanor.lib.gla.ac.uk/record=b1633964 [accessed 7 January 2013].
11. See C. Haines, 'Russell, Edward Stuart (1887–1954)', *Oxford Dictionary of National Biography* (Oxford: Oxford University Press, 2004).
12. See N. Roll-Hansen, 'E. S. Russell and J. H. Woodger: The Failure of Two Twentieth-Century Opponents of Mechanistic Biology', *Journal of the History of Biology*, 17:3 (1984), pp. 399–428.
13. Von Bertalanffy, though less so than Russell, gave several contributions to the journal, especially in the form of reviews.
14. In biology, C. M. Child, Y. Delage, E. Rabaud, C. L. Morgan, R. Semon, R. Goldschmidt, C. S. Sherrington, J. Needham and others were published in *Scientia*. For a general history of the journal, see S. Linguerri, *La Grande Festa della Scienza. Eugenio Rignano e Federico Enriques. Lettere* (Milan: Franco Angeli, 2005).
15. See E. V. Stonequist, 'Eugenio Rigano 1870–1930', *American Journal of Sociology*, 36:2 (1930), pp. 282–4; and G. Sarton, 'Eugenio Rignano', *ISIS*, 15:1 (1931), pp. 158–62.
16. F. Enriques, 'Motivi dell Filosofia di Eugenio Rignano', *Scientia*, 24:47 (1930), pp. 337–84.
17. E. Rignano, *Man Not a Machine: A Study of the Finalistic Aspects of Life*, with a foreword by H. Driesch (London: Kegan Paul, Trench, Trübner and Co., 1926).
18. J. Needham, *Man a Machine: In Answer to a Romantical and Unscientific Treatise Written by Sig. Eugenio Rignano & Entitled 'Man Not a Machine'* (London: Kegan Paul, Trench, Trübner and Co., 1928), p. 40. Haraway (*Crystals, Fabrics, and Fields*) and S. Winchester

(*The Man Who Loved China: The Fantastic Story of the Eccentric Scientist Who Unlocked the Mysteries of the Middle Kingdom* (New York: Harper Editions, 2008)) have described Needham as holist and organicist. However, according to the letters and published and unpublished materials I have consulted, I have noted that Needham was rather criticized by his organismic fellows. I can confidently say that at least during the early 1930s, he was not part of the organismic crowd, but instead critical of them. I think that Needham's bio-philosophy requires a serious rereading and rethinking, because even though he supposedly supported an organicist stance, it was a very peculiar and idiosyncratic stance.

19. E. Rignano, *Das Gedachtnis als Grundlage des Lebendigen; mit einer einleitung von Ludwig von Bertalanff* (Vienna: Wilhelm Braumuller, 1931).
20. E. S. Russell, 'Évolution ou épigénèse', *Scientia*, 8 (1910), pp. 225–46, on p. 225 (my translation from the French). Yet in 1912 Russell reviewed Rignano's English edition and concluded that 'We are convinced that the solution for the problem of heredity and development must be sought in the particular phenomena that Rignano, with a rare acquaintance, has chosen as specifically important, and we deeply recommend his work, which constitutes to our eyes the best attempt so far to produce a simpler and more rational solution in the framework of a positive philosophy'; E. S. Russell, 'Review of *Sur la Transmissibilité des Caractères Acquis: Hypothèse d'une Centro-épigénèse*', *Scientia*, 11 (1912), p. 439 (my translation from the French).
21. E. S. Russell, 'Vitalism', *Scientia*, 9 (1911), pp. 329–45.
22. Ibid., p. 330.
23. Ibid., p. 332.
24. See E. S. Russell, 'Fishery Researches: Its Contribution to Ecology', *Journal of Ecology*, 20:1 (1932), pp. 128–51.
25. Russell, 'Vitalism', p. 336.
26. Ibid.
27. Ibid., p. 373.
28. S. M. Persell, *Neo-Lamarckism and the Evolution Controversy in France, 1870–1920* (New York: Edwin Mellen Press, 1999), p. 34.
29. Ibid.
30. Russell, *Form and Function*, p. 45.
31. Ibid., p. 49.
32. Owen first introduced the distinction between analogical and homological organs which represented, Russell argued, a distinction between structure and function: indeed, function, as a teleological property, characterized analogical organs.
33. Russell, *Form and Function*, p. 145.
34. Ibid., p. 257. Russell here is referring to the law according to which the structures visible during the earlier stages of ontogeny are more common among all organisms than structures appearing in later ontogenic stages. This subject is treated in Richards, *The Meaning of Evolution*.
35. Russell, 'Vitalism', p. 343.
36. E. S. Russell, *The Study of Living Things: Prolegomena to a Functional Biology* (London: Methuen, 1924), p. vii.
37. Ibid., p. 46.
38. Ibid., pp. 12–13.
39. Bud, *The Uses of Life*.
40. Russell, *The Study*, p. 57.

41. Russell: 'it is mainly through perception that life becomes individualized and separates itself out from the environing flux. Through perception the organism clears, as it were, a space around it in which to live, and disposes of time in which to protect itself against the surrounding influences which would imminently destroy it' (ibid., p. 59).
42. Ibid., p. 61.
43. Russell reported several examples of creative behaviour among Protista and unicellular organisms.
44. Russell, in referring to the work of Kepner and Edwards on *Amoeba* and *Pelomyxa* feeding habits: W. A. Kepner and W. C. Whitlock, 'Food-Reactions of Pelomyxa carolinensis', *Journal of Experimental Zoology*, 24 (1917), pp. 381–404.
45. Bergson, *L'Évolution*, p. 35.
46. Russell, *The Study*, p. 77.
47. Thompson to Russell, 29 December 1916, letter ms14329, DTP-SAA.
48. Russell to Thompson, 6 January 1917, letter ms14333, DTP-SAA.
49. Russell, *The Study*, p. 132.
50. Russell, *The Interpretation of Development*, p. 82.
51. See E. S. Russell, 'Bateson, W. – Mendel's Principles of Heredity', *Scientia*, 15 (1914), pp. 274–8 (my translation from the French).
52. E. S. Russell, 'Vererbungslehre', *Scientia*, 15 (1914), pp. 278–81.
53. Russell, *The Interpretation of Development*, p. 8.
54. Ibid., p. 24.
55. The similarity between von Baer's and Child's conceptions of reproduction is striking.
56. Von Baer, quoted in Russell, *The Interpretation of Development*, p. 35.
57. Ibid., p. 67.
58. Ibid., p. 59.
59. F. B. Churchill, 'William Johannsen and the Genotype Concept', *Journal of the History of Biology*, 7:1 (1974), pp. 5–30.
60. W. Johannsen, 'Some Remarks about Units in Heredity', *Hereditas*, 4 (1923), pp. 133–41, on p. 136.
61. Ibid.
62. Ibid., p. 138.
63. Ibid., p. 139.
64. Russell, *The Interpretation of Development*, p. 31.
65. See A. *L'Œuf et les Facteurs de L'Ontogenèse* (Paris: G. Doin et Cie, 1917); and Mulnard, 'The Brussels School'.
66. As Russell claimed: 'It is these *special* resemblances and differences that have been the subject of the modern study of heredity, whether by biometrical or by genetical methods. The broad general resemblances of type give no hold for experimental or statistical treatment, and have accordingly on the whole been ignored. But it is this general hereditary resemblance which constitutes the main problem. We saw in discussing the gene theory that it deals only with differences between closely allied forms, and with the modes of inheritance of these differences; it leaves the main problem quite untouched as to why, for example, from a pair of *Drosophila* only *Drosophila* arise. It takes for granted the inheritance of Johannsen's "great central something" – the general hereditary equipment of the species' (Russell, *The Interpretation of Development*, p. 270).
67. In particular, F. R. Lillie, 'The Gene and the Ontogenetic Process', *Science*, 66:1712 (1927), pp. 361–8. In this paper Lillie had argued that gene theory could not help embryology and developmental theories.

68. Russell also included in his list Whitehead, who, as a philosopher, had provided the theoretical scaffolding for organismal theory.
69. Russell, *The Interpretation of Development*, p. 177.
70. Russell referred to Bernard's *La Science Experimentale*, a book published in 1878, where Bernard effectively admitted: 'In saying that life is a directive idea or an evolutive force of being, we simply convey the idea that there is an unity in all morphological and chemical changes that are produced, at beginning, by the germ until the end of life'; C. Bernard, *La Science Experimentale* (Paris: Bailliere, 1878), p. 430 (my translation from French).
71. As Russell specified: 'All parts and organs are, as Kant would say, reciprocally means and ends, and all cooperate in the life of the organism as a whole. For, as D'Arcy Thompson, "the life of the body is more than the sum of the properties of the cells of which it is composed"' (Russell, *The Interpretation of Development*, p. 148).
72. Ibid., pp. 92–3. As Russell emphasized: 'This principle of unity, or action of the organism as a whole, corresponds to Whitman's concept of organization ... the same view, that development is essentially an activity of the organism as a whole, has also been upheld by others – by Conklin and Child for instance, and by Ritter' (ibid., p. 244).
73. Ibid., p. 189.
74. Roll-Hansen, 'E. S. Russell and J. H. Woodger', p. 409.
75. Von Bertalanffy, *Modern Theories*, p. vii.
76. Russell, *The Interpretation of Development*, p. 192.
77. Roll-Hansen, 'E. S. Russell and J. H. Woodger', p. 410.
78. Russell, *The Study*, pp. x–xi.
79. Whitehead, *Science*, p. 21.
80. J. H. Woodger, *Biological Principles: A Critical Study* (London: Kegan Paul, Trench, Trübner and Co., 1929), p. i.
81. Woodger quoted Cassirer from *Einstein's Theory of Relativity Considered from the Epistemological Standpoint* (Chicago, IL: Open Court, 1923). The reference to Cassirer was very indicative; indeed, Cassirer investigated the connection between Einstein's relativity and Kant's theory of knowledge.
82. Woodger, *Biological Principles*, p. 6.
83. Woodger explicitly mentioned that his method recalled Kant's approach to philosophical investigation: 'Criticism as here intended (in the sense in which it was first introduced into philosophy by Kant) is a disinterested examination of traditional conflicts with a view to the discovery of their roots, and the removal of difficulties created by an uncritical use of the notion of unreflective thought' (ibid., p. 7).
84. Roll-Hansen, 'E. S. Russell and J. H. Woodger', p. 408.
85. See, for instance, J. H. Woodger, *The Axiomatic Method in Biology* (Cambridge: Cambridge University Press, 1937).
86. J. Cain, 'Woodger, Positivism and Evolutionary Synthesis', *Biology and Philosophy*, 15:4 (2000), pp. 535–51, on p. 539
87. See V. B. Smocovitis, 'Unifying Biology: The Evolutionary Synthesis and Evolutionary Biology', *Journal of the History of Biology*, 25 (1992), pp. 1–65; and V. B. Smocovitis, *Unifying Biology: The Evolutionary Synthesis and Evolutionary Biology* (Princeton, NJ: Princeton University Press, 1996).
88. Cain, 'Woodger', p. 541.
89. J. H. Woodger, *Elementary Morphology and Physiology for Medical Students: A Guide for the First Year and a Stepping-Stone to the Second* (Oxford: Oxford University Press, 1924).

90. See D'Arcy Thompson to Woodger, 29 November 1925, Box C1/3, Folder: Correspondence D'Arcy Thompson, Woodger Papers, in Special Collections, University College London (hereafter UCL Archives); and C. L. Edwards, 'The Vienna Institution for Experimental Biology', *Popular Science Monthly*, 78:37 (1911), pp. 584–601
91. See J. H. Woodger 6 January 1930, Academic Registrar, University of London, C 1/3, Miscellaneous, Woodger Papers, UCL Archives. For a biographical sketch, see also Gregg and Harris, *Form and Strategy*.
92. The first papers that Woodger published were rather technical and essentially descriptive. See: 'On the Relationship between the Formation of Yolk and the Mitochondria and Golgi Apparatus during Oogenesis', *Journal of the Royal Microscopical* Society, 40:2 (1920), pp. 129–56; 'Notes on a Cestode Occurring in the Haemocoele of House-Flies in Mesopotamia', *Annals of Applied Biology*, 3 (1921), pp. 346–51; On the Origin of the Golgi Apparatus on the Middle-Piece of the Ripe Sperm of Cavia', *Quarterly Journal of Microscopal Science*, 65 (1921), pp. 265–91; and 'Observations on the Origin of the Germ-Cells in the Fowl, Studied by Means of their Golgi Bodies', *Quarterly Journal of Microscopal Science*, 69 (1925), pp. 445–62.
93. On the Vienna Vivarium, see Cohen, 'Living Precisely'.
94. See Edwards, 'The Vienna Institution'.
95. The father of Wolfgang Pauly (1900–58), Nobel Laureate in Physics.
96. Przibram, quoted in Drack et al., 'On the Making', p. 357.
97. D. Pouvreau, *Une Biographie Non Officielle de Ludwig von Bertalanffy (1901–1972)* (Vienna: Bertalanffy Center for the Study of Systems Science, 2006), at http://www.bertalanffy.org/media/pdf/pdf29.pdf [accessed 13 August 2013], p. 13 (my translation from French).
98. Przibram, quoted in Drack et al., 'On the Making', p. 354.
99. See Koestler, *The Case*. See also S. Gliboff, 'The Case of Paul Kammerer: Evolution and Experimentation in the Early 20th Century', *Journal of the History of Biology*, 39:3 (2006), pp. 525–63.
100. See J. Overton, 'Paul Alfred Weiss 1898–1989', in *Biographical Memoirs: National Academy of Sciences* (Washington, DC: National Academies Press, 1997), at http://www.nap.edu/readingroom.php?book=biomems&page=pweiss.html [accessed 13 August 2013]; and S. Brauckmann, 'The Scientific Life of Paul A. Weiss (1898–1989)', *Mendel Newsletter*, 12 (2003), pp. 2–7.
101. On the intellectual relations and contacts between Weiss, von Bertalanffy and the Vienna Vivarium, see Drack et al., 'On the Making'. On von Bertalanffy's PhD dissertation and life, see Pouvreau, *Une Biographie Non Officielle de Ludwig von Bertalanffy*.
102. Przibram himself, as expressed in a letter of recommendation, praised Woodger's achievements. See Przibram, 14 December 1929, Box C1/3, Folder: Miscellaneous, Woodger Papers, UCL Archives.
103. Woodger, *Biological Principles*, p. 291.
104. See Gregg and Harris, *Form and Strategy*, p. 2.
105. Woodger, *Elementary Morphology*, p. 2.
106. Ibid., pp. 457–8.
107. Woodger, *Biological Principles*, p. 273.
108. Ibid., p. 275.
109. Ibid., p. 293.
110. Ibid., p. 311.
111. As Roll-Hansen and Ruse have argued; see Roll-Hansen, 'E. S. Russell and J. H. Woodger', and M. Ruse, 'Woodger on Genetics, a Critical Evaluation', *Acta Biotheoretica*, 24 (1975), pp. 1–13.

112. See R. Pearl's letter of recommendation to Woodger, Box C1/3, Folder: Miscellaneous, Woodger Papers, UCL Archives.
113. J. H. Woodger, 'The Concept of Organism and the Relation between Embryology and Genetics Part II', *Quarterly Review of Biology*, 5:4 (1931), pp. 438–62, on p. 452.
114. J. H. Woodger, 'The Concept of Organism and the Relation between Embryology and Genetics Part III', *Quarterly Review of Biology*, 6:2 (1931), pp. 178–207, on p. 180.
115. Woodger, *Biological Principles*, p. 384.
116. J. H. Woodger, 'The Concept of Organism and the Relation between Embryology and Genetics Part I', *Quarterly Review of Biology*, 5:1 (1930), pp. 1–22, on p. 15.
117. Ibid., p. 18.
118. Apparently, when Woodger arrived in Vienna, the worms he found there were not appropriate for transplantation experiments. See Gregg and Harris, *Form and Strategy*, p. 3.
119. H. Przibram, 'Regeneration und Transplantation im Tierreich', *Kultur der Gegenwart, Band Allgemeine Biologie*, (1913), pp. 347–77. The draft translation is in Woodger Papers, Box D4, Folder: Embryology, p. 17, UCL Archives.
120. On Weiss's experiments, see Haraway, *Crystals, Fabrics, and Fields*.
121. Woodger, *Biological Principles*, p. 355.
122. On Spemann's experiments, see V. Hamburger, *The Heritage of Experimental Embryology, Hans Spemann and the Organizer* (Oxford: Oxford University Press, 1988); see also H. Spemann, 'Experimentelle Forschungen zum Determinations und Individualitäts Problem', *Naturwissenschaften*, 7 (1919), pp. 581–91.
123. In 1926 T. H. Morgan had published a well-known paper in the *American Naturalist* arguing, among other things, that genes, if considered as protein bodies, could produce enzymes and therefore 'affect' development; see T. H. Morgan, 'Genetics and Physiology of Development', *American Naturalist*, 60:671 (1926), pp. 489–515.
124. Woodger, *Biological Principles*, p. 376.
125. Ibid., pp. 365–6.
126. Ibid., p. 205.
127. See Woodger, 'The Concept of Organism Part III'.
128. Ibid., p. 206.
129. See P. Abir-Am, 'The Biotheoretical Gathering, Trans-Disciplinary Authority and the Incipient Legitimating of Molecular Biology in the 1930s: New Perspective on the Historical Sociology of Science', *History of Science*, 25:67(1) (1987), pp. 1–70.
130. Werskey defines the Theoretical Biology Club as 'one of the most important underground scientific enterprises of the 1930s'; Werskey, *The Visible College*, p. 80.
131. Woodger, *Biological Principles*, p. 66.
132. See Woodger, 'A Biological Approach to Socialism' [typescript], *c.* 1945, Woodger Papers, Box E/3, UCL Archives.
133. Woodger once again used the analogy with animal regeneration and transplantation; see ibid., p. 5.
134. See ibid., p. 6.
135. See ibid., p. 9.
136. See Werskey, *The Visible College*, p. 138.

## 5 The American Version: Chicago and Beyond

1. See E. Lurie, *Louis Agassiz: A Life in Science* (Chicago, IL: University of Chicago Press, 1960).
2. Winsor, *Reading the Shape*; and P. J. Pauly, *Biologists and the Promise of American Life: From Meriwether Lewis to Alfred Kinsey* (Princeton, NJ: Princeton University Press, 2000).

3. Among others, J. Le Conte, A. Hyatt, D. S. Jordan and A. P. Packard.
4. See K. R. Benson, 'Why American Marine Stations?: The Teaching Argument', *American Zoologist*, 28:1 (1988), pp. 7–14; and J. Maienschein, 'Agassiz, Hyatt, Whitman and the Birth of Marine Biological Laboratory', *Biological Bulletin*, 168 (1985), pp. 26–34. See also R. Rainger, K. R. Benson and J. Maienschein (eds), *The American Development of Biology* (New Brunswick, NJ: Rutgers University Press, 1991).
5. See W. O. Hangstrom, *The Scientific Community* (Carbondale, IL: Southern Illinois University Press, 1975); and J. Ben-David and T. A. Sullivan, 'Sociology of Science', *Annual Review of Sociology*, 1 (1975), pp. 203–22.
6. J. Harwood, *Styles of Scientific Thought: The German Genetics Community, 1900–33* (Chicago, IL: University of Chicago Press, 2003), p. 177.
7. Ibid., p. 178
8. Ibid., p. 156.
9. See R. Hofstadter, *Anti-Intellectualism in American Life* (New York: Random House, 1963).
10. Harwood, *Styles*, p. 190.
11. I mean all models that regard the development of scientific ideas as mere epiphenomena of institutional or social organizations, structures of power, economical constraints or interests, or 'basic intentions'. That does not mean I consider all externalist approaches unhelpful or useless; I only question the idea that any tradition, style of thought or theoretical result has a direct and reducible relation with its social and institutional environment.
12. See H. H. Newmann, 'History of the Department of Zoology in the University of Chicago', *Bios*, 19:4 (1948), pp. 215–39.
13. Morse, *Biographical Memoir of Charles Otis Whitman*.
14. G. Mitman, *The State of Nature: Ecology, Community, and American Social Thought, 1900–1950* (Chicago, IL: University of Chicago Press, 1992), p. 22.
15. Pauly too stressed this point: 'his organicist belief that biologists like, organisms, would interact to produce a whole larger than the sum of the parts and would thereby individually become enriched. Success, while natural, would not be automatic. It depended on a favorable environment, the right people, intimate associations, and continued autonomy' (Pauly, *Biologists*, pp. 153–4).
16. Lillie had earned his PhD in zoology at the same institution.
17. J. Maienschein, 'Whitman at Chicago: Establishing a Chicago Style of Biology?', in R. Rainger, K. R. Benson and J. Maienschein (eds), *The American Development of Biology* (Philadelphia, PA: University of Pennsylvania Press, 1991), pp. 151–80.
18. F. R. Lillie, 'The Organization of the Egg in *Unio*, Based on a Study on its Maturation, Fertilization, and Cleavage', *Journal of Morphology*, 17 (1901), pp. 277–92.
19. F. R. Lillie, 'Observations and Experiments Concerning the Elementary Phenomena of Embryonic Development in *Chaetopterus*', *Journal of Experimental Zoology*, 3 (1906), pp. 153–268.
20. See Willier, 'F. R. Lillie'. See also Lillie, 'My Early Life' [a biography written for Lillie's children in 1944], Lillie Papers, Box II I, Folder 8, MBL Archive, Woods Hole.
21. See Lillie Obituary, 1947, Lillie Papers, Box II I, Folder 7, MBL Archive, Woods Hole.
22. See F. R. Lillie, *The Woods Hole Marine Biological Laboratory* (Chicago, IL: University of Chicago Press, 1944).
23. Newman, 'History of the Department of Zoology', p. 229; see also P. Peters, 'Frank Lillie: The Cultivator of an Embryonic Woods Hole Scientific Community', *Cape Cod Times*, 6

June 1999, at http://www.capecodonline.com/apps/pbcs.dll/article?AID=/19990606/NEWS01/306069977&cid=sitesearch&template=printart [accessed 13 August 2013].
24. K. R. Manning, *Black Apollo of Science: The Life of Ernest Everett Just* (Oxford: Oxford University Press, 1983).
25. J. Farley, *Gametes and Spores, Ideas about Sexual Reproduction: 1750–1914* (Baltimore, MD: Johns Hopkins University Press, 1982).
26. F. R. Lillie, 'Studies on Fertilization: VI. The Mechanism of Fertilization in *Arbacia*', *Journal of Experimental Zoology*, 16 (1914), pp. 523–90.
27. Lillie, 'Observations and Experiments', p. 154.
28. Ibid., p. 246.
29. Ibid., p. 250.
30. W. Johannsen, *Elemente der exakten Erblichkeitslehere* (Jena: Gustav Fisher, 1909); and W. Johanssen, 'The Genotype Conception of Heredity', *American Naturalist*, 45 (1911), pp. 129–59. See also Churchill, 'William Johannsen'.
31. Lillie, 'Observations and Experiments', p. 251.
32. This last point will be developed further on.
33. Lillie, 'Observations and Experiments', p. 252; Whitman had pointed out such a hypothesis in his article 'The Inadequacy of the Cellular Theory of Development' (1893).
34. Lillie, 'Observations and Experiments', p. 251.
35. Ibid., p. 252.
36. Lillie distinguished between the 'action of the whole organism', in which the entire organism 'exercises a formative influence on all of its parts', and 'correlative differentiation', which involved 'all actions of the intraorganic environment'. The action of the whole organism on its parts was, after all, a qualitative extension of the more limited correlative differentiations.
37. Lillie, 'Observations and Experiments', p. 258.
38. Ibid., p. 260.
39. Ibid.
40. Potencies and latencies as dominant and recessive.
41. F. R. Lillie, 'The Theory of Individual Development', *Popular Science Monthly*, 75:14 (1914), pp. 239–52.
42. As Lillie instantiated: 'the constancy of distribution of peripheral nerve is not due to the transmission of nerve-branching determinants from generation to generation, but is a function of the intra-organic environment in each generation' (ibid., p. 244).
43. Ibid., p. 245.
44. Ibid., p. 248.
45. Again, 'characters' do not represent static and definite entities because 'in the study of heredity and development we are dealing with biological processes' (ibid., p. 250).
46. Ibid.
47. Ibid.
48. Ibid., p. 252.
49. See G. E. Allen, 'Heredity under an Embryological Paradigm: The Case of Genetics and Embryology', *Biological Bulletin*, 68 (1985), pp. 107–21.
50. See Sapp, *Beyond the Gene*; and F. S. Gilbert, 'Bearing Crosses: The Historiography of Genetics and Embryology', *American Journal of Medical Genetics*, 76:2 (1998), pp. 168–82.
51. Lillie, 'The Gene'.
52. Lillie mentioned Morgan, Spemann and Goldschmidt.

53. Lillie accepted Child's theory of metabolic gradients.
54. This will be discussed more extensively in the next section.
55. Lillie would extend this model in another paper published in 1929, which I will discuss later on.
56. Lillie, 'The Gene', p. 366.
57. Lillie in particular criticized Goldschmidt's attempts to formulate a physiological theory of development positing a complex mechanism of gene activations at different times. Goldschmidt's model indeed 'presupposes an underlying mechanism adequate to almost ultraphysically precise regulation in the germ, for which no model can be suggested: at the most it shifts the difficulty one step further back' (ibid., p. 365).
58. Ibid., p. 366.
59. We know that Morgan clearly admitted this last point: 'Between the character (phenotype), that furnish the data for the theory and the postulated genes, to which the characters are referred; lies the whole field of embryonic development. The theory of the gene, as here formulated states nothing with respect to the way in which the genes are connected with the end-product, or character'; T. H. Morgan, *The Theory of the Gene* (New Haven, CT: Yale University Press, 1926), p. 26.
60. Lillie, 'The Gene', p. 368.
61. F. R. Lillie and R. Juhn, 'The Physiology of Development of Feathers. I. Growth-Rate and Pattern in the Individual Feather', *Physiological Zoology*, 5:1 (1932), pp. 124–84; then F. R. Lillie, 'Zoological Sciences in the Future', *Science*, 88:2273 (1938), pp. 65–72.
62. F. R. Lillie, 'Physiology of Development of the Feather. II. General Principles of Development with Special Reference to the After-Feather', *Physiological Zoology*, 11:4 (1938), pp. 434–50.
63. F. R. Lillie, 'The Physiology of Feather Pattern', *Wilson Bulletin*, 44:4 (1932), pp. 193–211.
64. F. R. Lillie, 'Physiology of Development of the Feather. V. Experimental Morphogenesis', *Physiological Zoology*, 14:2 (1941), pp. 103–35; and F. R. Lillie, 'On the Development of Feathers', *Biological Reviews*, 17:3 (1942), pp. 247–66.
65. Lillie and Juhn, 'The Physiology of Feathers I', pp. 177–8.
66. Ibid., p. 175.
67. Lillie, 'The Physiology of Feather Pattern', p. 207.
68. As we have seen with the paper Lillie published in 1909, colour variation in mammals, as well as phenomena related to regeneration, the anatomy of the nervous system, development of the blood vessels, etc. did not depend on the existence of determinants or definite unit-factors, but on dynamic intra- and extra-organic interactions.
69. Lillie, quoted in J. Sapp, *Genesis: The Evolution of Biology* (Oxford: Oxford University Press, 2003), p. 136.
70. As Sapp clearly explains: 'The notion that the gene determines the characteristic was pure tautology in the typical breeding experiments because the presence of genes was inferred by experimental manipulation of phenotypes' (ibid., p. 136).
71. See F. B. Churchill, 'From Machine-Theory to Entelechy: Two Studies in Developmental Teleology', *Journal of the History of Biology*, 2:1 (1969), pp. 165–85.
72. Lillie, 'Embryonic Segregation and its Role in the Life History', *Development Genes and Evolution*, 118:1 (1929), pp. 499–533, on p. 500.
73. Ibid., p. 505.
74. However, Lillie warned that the analogy between dominant and recessive genes and negative and positive determined segregates must not be pushed too far: they were two

different phenomena. Indeed, the processes related to embryonic segregation concerned the cytoplasm, whereas the Mendelian mechanisms referred to the cell nucleus.

75. Lillie, 'Embryonic Segregation', pp. 516–28.
76. Sapp, *Beyond the Gene*, p. 7.
77. Just, on the other hand, as we will see in the next sections, regarded Lillie's theory as an alternative to genetics.
78. We have seen that Lillie considered genetics the science of phenotypes.
79. Lillie, 'Embryonic Segregation', p. 507.
80. Lillie, quoted in Mitman, *The State*, p. 98–9.
81. Ibid., p. 100.
82. On the history of this institute, see ibid., pp. 96–109. See also Kay, *The Molecular Vision*.
83. See M. Morange, *A History of Molecular Biology* (Cambridge, MA: Harvard University Press, 2000).
84. Lillie to Ritter, 16 November 1938, Box IIA, Folder 76, 1, Lillie Papers, MBL Archive, Woods Hole.
85. Lillie, 'Physiology of the Feather II', p. 67.
86. 'Nature has neither kernel nor shell. She is both at the same time' (my translation); Goethe, quoted in E. E. Just, *The Biology of the Cell Surface* (London: Technical Press, 1939), p. 1.
87. Lillie to Harrison, 20 March 1939, Box 6, Folder II A 54, MBL Archive, Woods Hole.
88. With the exception of F. S. Gilbert's excellent article 'Cellular Politics: Ernest Everett Just, Richard B. Goldschmidt, and the Attempt to Reconcile Embryology and Genetics', in R. Rainger, K. R. Benson and J. Maienschein (eds), *The American Development of Biology* (Philadelphia, PA: University of Pennsylvania Press, 1991), pp. 311–46, in which he compares Just's and *Goldschmidt's* theoretical alternatives to gene-centric hypotheses on heredity and development.
89. Just's father was the son of a slave.
90. Manning, *Black Apollo*, p. 85.
91. In 1936 Just even made a formal request to Mussolini in order to get founded in Italy; see ibid., p. 291.
92. With the exception of a recent article written by Jan Sapp – 'Just in Time: Gene Theory and the Biology of the Cell Biology', *Molecular Reproduction and Development*, 76 (2009), pp. 1–9 – which briefly tackles Just's biology, all the other contributions are mostly popularizations of the individual, and not his science. See S. J. Gould, 'Thwarted Genius', *New York Review of Books*, 30:18 (24 November 1983), at http://www.nybooks.com/articles/archives/1983/nov/24/thwarted-genius/ [accessed 13 August 2013]; S. J. Gould, 'Just in the Middle: A Solution to the Mechanist-Vitalist Controversy', in *The Flamingo's Smile: Reflections in Natural History* (New York: W. W. Norton, 1987), pp. 337–91; W. M. Byrnes, 'Ernest Everett Just, Johannes Holtfreter, and the Origin of Certain Concepts in Embryo Morphogenesis', *Molecular Reproduction and Development*, 76 (2009), pp. 912–21; and W. M. Byrnes and W. R. Eckberg, 'Ernest Everett Just (1883–1941) – An Early Ecological Developmental Biologist', *Developmental Biology*, 296 (2006), pp. 1–11.
93. Since 1927 he has been included in the *American Men of Science*'s list of the top thirty-eight zoologists in America. In 2002 the black American historian Molefi Kete Asante, a specialist in African American studies, included Just in the list of 100 greatest African Americans.
94. Just, *Biology*.
95. Manning, *Black Apollo*.

96. Lillie to W. Weaver, 6 March 1933, Box 6, Folder II A 54 (Just 1912–41), p. 1, Lillie Papers, MBL Archive, Woods Hole.
97. Ibid, p. 2.
98. E. E. Just, 'On the Origins of Mutations', *American Naturalist*, 66 (1932), pp. 61–74.
99. Ibid., p. 69.
100. As Just specified in the same article: 'The cell is a unit: the nucleus influences the plasma, the plasma the nucleus. The cell reacts as a whole' (ibid., p. 73).
101. Ibid., p. 74.
102. Ibid.
103. E. E. Just, 'Cortical Cytoplasm and Evolution', *American Naturalist*, 66 (1933), pp. 22–9.
104. In 1913 L. J. Henderson had published an influential book: *The Fitness of the Environment* (New York: Macmillan, 1913). See also J. Parascandola, 'Organismic and Holistic Concepts in the Thought of L. J. Henderson', *Journal of the History of Biology*, 4:1 (1971), pp. 63–113.
105. Just, 'Cortical Cytoplasm', p. 23.
106. We have seen that D'Arcy Thompson too propounded such a hypothesis; see Chapter 3 in this volume.
107. Just, 'Cortical Cytoplasm', p. 26.
108. Ibid., p. 28.
109. See E. E. Just, 'A Single Theory for the Physiology of Development and Genetics', *American Naturalist*, 70 (1936), pp. 267–312; and E. E. Just, 'Phenomena of Embryogenesis and their Significance for a Theory of Development and Heredity', *American Naturalist*, 71 (1937), pp. 97–112.
110. Just, 'A Single Theory', p. 272.
111. A hypothesis advocated, among others, by Morgan himself.
112. Just, 'A Single Theory', pp. 280–1.
113. Just believed that regeneration happened when environmental factors, such as an injury, fostered the release of material from nucleus to cytoplasm; all the potencies that had been taken off from the cytoplasm during development. Tumours were explainable according to the same line of reasoning; they were due to a sudden releasing of nuclear potencies once they were removed from cytoplasm.
114. Just, 'A Single Theory', p. 290. Notice the language. Just, like many of his contemporaries, deemed gene theory only one theory of heredity among others.
115. Ibid., p. 298.
116. Just, 'Phenomena of Embryogenesis', p. 108.
117. Just, 'A Single Theory', p. 305.
118. See also Gilbert, 'Cellular Politics'.
119. In the same year that his monograph was published, Just launched a new research project that he had described in short typescript, entitled 'Status of my Research Program in Embryology and its Implication for General Biology' – a document that Just sent to Lillie. The central question of this project was, of course, how a single cell can form a whole organism, a question that, if answered – Just argued – could not only shed light on embryological issues, but also provide fundamental information to general biology and medicine. The project included a detailed list of all the phenomena requiring further investigation: from fertilization and cytoplasmic cleavage to the study of enucleated egg development and egg axis formation. See E. E. Just, 1912–41, Box 6, Folder II A 54, 6, Lillie Papers, MBL Archive, Woods Hole.

120. Lillie really appreciated Just's book; in a letter sent to Harrison, he wrote: 'I have just been through his new book [Just] and am quite impressed with his point of view with reference to the principles of scientific analysis in biology stated in the introduction, and with which I agree. He has stated the situation very well'; Lillie to R. G. Harrison, 20 March 1939, Box 6, Folder II A 54, MBL Archive, Woods Hole.
121. Just, *Biology*.
122. Ibid., p. 7.
123. Just was part of the tradition of *New Natural History* as Robert E. Kohler has defined it. Like many of his contemporaries, Just attempted to find a middle ground between results obtained in the lab and fieldwork observations. To him, both practices had to work together. Of course, this was probably the result of his long training at Woods Hole, where both lab experiment and observations in the field were daily practices. As Kohler well described the 'spirit' of the 'new naturalist': 'The ideal of the new naturalists was a synthesis of sympathy and intellect, observation and experiment, spontaneity and control, laboratory and field'; R. E. Kohler, *Landscapes and Labscapes: Exploring the Lab-Field Border in Biology* (Chicago, IL: University of Chicago Press, 2002), p. 34.
124. See also E. E. Just, 'Unsolved Problems of General Biology', *Physiological Zoology*, 13:2 (1940), pp. 123–42.
125. The ectoplasm is defined as the outer level of cytoplasm.
126. Just, *Biology*, p. 361.

## 6 Romantic Biology from California's Shores: W. E. Ritter, C. M. Child and the Scripps Marine Association

1. See J. Maienschein, *Transforming Traditions in American Biology, 1880–1915* (Baltimore, MD: Johns Hopkins University Press, 1991).
2. E. B. Wilson, *The Cell in Development and Inheritance* (London: Macmillan, 1896).
3. Ibid., p. 393.
4. See J. Maienschein, 'H. N. Martin and W. K. Brooks: Exemplars for American Biology?', *American Zoologist*, 27:3 (1987), pp. 773–83.
5. See E. S. Morse, *Biographical Memoir of Charles Sedgwick Minot 1852–1914* (Washington, DC: National Academy of Sciences, 1920).
6. On the American tradition of biology, see Pauly, *Biologists*.
7. Ritter, Diary 1894, 71/3c, 1883–95, Ctn-9, Bancroft Archive.
8. Ritter's short experience in Naples is reported on his diaries, now conserved in the Bancroft Library at the University of California, Berkeley.
9. Ritter to Dohrn, 12 July 1909, Dohrn Correspondence, Letter n. 186, Historical Archives, Stazione Zoologica Anton Dohrn (hereafter SZAD Archive).
10. M. T. Ghiselin, and C. Groeben, 'A Bioeconomic Perspective on the Organization of the Naples Zoological Station', in M. T. Ghiselin and A. E. Leviton (eds), *Cultures and Institutions of Natural History: Essays in the History and Philosophy of Science* (San Francisco, CA: California Academy of Sciences, 2000), pp. 273–85, on p. 274.
11. On Ritter's interest in Kant's *Critique of Judgment*, see below, p. 222 n. 61.
12. C. McWilliams, *Southern California: An Island on the Land* (Layton, UT: Gibbs Smith, 1973), p. 113.
13. Ibid.

14. See R. V. Hine, *California's Utopian Colonies* (San Marino, CA: Huntington Library, 1953).
15. See Pauly, *Biologists.*
16. On Scripps, see B. Kovarik, *Revolutions in Communication: Media History from Gutenberg to the Digital Age* (New York: Continuum, 2001).
17. See McWilliams, *Southern California.*
18. See L. D. Stephens, 'Joseph Leconte's Evolutional Idealism: A Lamarckian View of Cultural History', *Journal of the History of Ideas*, 39:3 (1978), pp. 465–80.
19. Herdman was a close friend of the holist physiologist J. S. Haldane and the zoologist D'Arcy W. Thompson; see Thompson, *D. W. Thompson.*
20. See 71/3c, Box-10, Biographical Record Folder: Ritter, 1935, Supplementary Biographical Record, for the Year 1935–6. Autobiographical Sketch: The Manner of the Man, Bancroft Archive.
21. T. Litwin, *The Harriman Alaska Expedition Retraced: A Century of Change, 1899–2001* (New Brunswick, NJ: Rutgers University Press, 2005).
22. See B. L. Norcross, 'Beyond Serenity: Boom and Bust in Prince William Sound', in T. Litwin (ed.), *The 1899 Harriman Alaska: A Century of Change* (New Brunswick, NJ: Rutgers University Press, 2005), pp. 115–24.
23. For a detailed history of the first fifty years of the Scripps Institution, see H. Raitt and B. Moulton, *Scripps Institution of Oceanography: First Fifty Years* (Los Angeles, CA: Ward Ritchie Press, 1966).
24. See E. N. Shor, 'How Scripps Institution Came to San Diego', *Journal of San Diego History*, 27:3 (1981), pp. 161–73; and W. E. Ritter, *The Marine Biological Station of San Diego, its History, Present Conditions, Achievements, and Aims* (Berkeley, CA: University of California Press, 1912).
25. See Ritter, 1915, Box-4, Folder: The Philosophy of E. W. Scripps, Historical Archives, Scripps Institution of Oceanography, University of California, San Diego (hereafter SIO Archive).
26. Ibid.
27. E. W. Scripps, 'The Scripps Institute for Biological Research', 15 September 1916, RP, Box-2, Folder-95, p. 9, SIO Archive.
28. W. E. Ritter, 'Organization in Scientific Research', *Popular Science Monthly*, 67 (1904), pp. 49–53.
29. As Pauly recounts: 'Scripps saw in Ritter's vision of biological research and theory some basis for understanding humans as political animals. He wanted biologists to provide Americans a naturalistic account of customs, ethics, religion, and philosophy that would increase their happiness and reinforce democracy' (Pauly, *Biologists*, p. 206).
30. E. W. Scripps, 'The Scripps Institute for Biological Research', 15 September 1916, RP, Box-2, Folder-95, p. 7, SIO Archive.
31. See M. H. Stevens, 'The Enigma of Meyer Lissner: Los Angeles's Progressive Boss', *Journal of the Gilded Age and Progressive Era*, 8:1 (2009), pp. 111–36.
32. Ritter, 24 August 1912, Meyer Lissner, RP, Box-15, Ctn-33, Folder: Outgoing Letters, 1912–14, Bancroft Archive.
33. In a letter addressed to the president of the University of California in 1912, Ritter wrote: 'Politically I am heart and head for the creed of the Progressive Party, and would like to show its leaders and others what, to me at least, its scientific meaning is ... never again can political doctrine safely ignore certain basal biological principles'; Ritter to Wheeler, 1912, RP, Box-1, Folder: Correspondence June–Aug, SIO Archive.

34. Ritter to Scripps, 1921, RP, 71/3c, Ctn. 33, Folder: Scripps Correspondence, Bancroft Archive.
35. Ritter stresses such a point in his letters quite often. In 1908 he even expressed his intention to tie together biology and human welfare in a letter to F. R. Lillie: 'I believe that within a limited period, say of five years more or less, the Station may be made one of the most influential biological foundations in existence ... I believe that the San Diego Station is in position to become an institutional center for this process of regeneration'; Ritter to Lillie, 25 March 1908, RP, Box-1, Folder: Correspondence-1908, SIO Archive.
36. As Ritter specified in an unpublished document addressed to the community of San Diego; see Ritter, 1910, RP, Box-3, Folder: Descriptions of the Scripps Institute of Oceanography, SIO Archive.
37. As Kohler describes the frontier between lab and field: 'The dynamics of the lab-field border began to change around the turn of the century, and in the following decades an uneasy frontier gradually evolved into a zone of interaction and mixed border practices. In the United States this change was marked initially by a spate of public pronouncements by leading biologists that the laboratory movement had gone too far and that it would be to all biologists' advantage to combine the older natural history with the more recent experimental tradition' (Kohler, *Landscapes*, p. 24).
38. Ritter to Barrow, 1922, RP, 71/3c, Ctn. 33, Folder: Barrows, David Prescott, Bancroft Archive.
39. See W. B. Provine, 'Francis B. Sumner and the Evolutionary Synthesis', *Studies in History of Biology*, 3 (1979), pp. 211–40.
40. In particular, for Sumner, field and laboratory work undertaken on mice proved that small inherited variations were due to natural selection. As Koehler argues, the results that Sumner obtained were related to his unorthodox approach to studying genetics in the field: 'Sumner carried on experimental breeding at the Scripps mouse colony or "murarium", crossing subspecies to see how species characters were inherited. The results agreed against mutations, he believed, despite the growing consensus among lab geneticists that "blending" inheritance was the result of normal Mendelian segregation in multiple-factor characters. Sumner's genetic apostasy was the result of applying methods designed for standard animals to the complex traits of a wild species – the result, in other words, of bringing nature into the lab'(Kohler, *Landscapes*, p. 143).
41. On Sumner's agenda on heredity, see C. M. Child, *Biographical Memoir of Francis Bertody Sumner, 1874–1945* (Washington, DC: National Academy of Science, 1946), pp. 12–13.
42. See Survey of Scripps Institution for Biological Research, Box-3, Folder: Survey of SIO for Bio. Research, SIO Archive.
43. In November 1919 Ritter sent two letters to Sumner, in which he advised him to consider a possible line of research that entailed a study of the relation of the human races. Ritter and Scripps considered such a topic a hot issue of the time, especially because they related it to American immigration policies. On these discussions, see Ritter to Sumner, 13 and 17 November 1919, RP, Box-3, Folder: Correspondence Sept–Dec 1919, SIO Archive.
44. Ritter endorsed a physiological theory of heredity in which environment played a central role. In fact, Ritter did not believe there was a neat separation between nature and nurture, and this surely reinforced his scepticism about eugenic policies. In a letter he sent to Irving Fisher in 1922 (see RP, 71/3c, Ctn. 33, Folder: Fisher Irving, 1867–1947, Bancroft Archive) he harshly criticized eugenic agendas; he believed that such a social programme was doomed to failure because it was based on a mistaken conception of heredity, a con-

ception that overlooked development. He concluded that eugenics, as it was based on Weismannian biology, was undemocratic, amoral and in contradiction with a liberal state. On the other hand, he believed that race improvement depended on the betterment of the social environment. Children's education and cultural environment was central for him.

45. F. E. A. Thone and E. W. Baley, 'William Emerson Ritter: Builder', *Scientific Monthly*, 24 (1927), pp. 256–62, on pp. 259–260.
46. Ritter to B. I. Wheeler, 24 August, 1912, Box-1, Folder Correspondence June–Aug, SIO Archive.
47. Ibid., p. 1.
48. Ritter to Merritt, 1 August 1914, Box-2, Folder Correspondence Apr–Sept, SIO Archive, p. 1.
49. Ibid., p. 2.
50. Ibid., p. 3.
51. E. H. Harriman financed the Alaska exploration, in which Ritter, as I have already mentioned, participated as zoologist.
52. 'For one I am satisfied the time has come to rebuild the whole theory of living things, man included, as thoroughly as your father and his associates rebuilt the rail-road systems which came into their hands. I further believe that we here at the San Diego Station are in position to play a large part in this reconstruction'; letter to Harriman, 8 February 1910, RP, Ctn.1, Folder: Harriman Outgoing Letters, 1910–11, Bancroft Archive.
53. Ritter to Scripps, 24 October 1921, RP, Box-1, Folder-10, SIO Archive.
54. Ritter to Scripps, 4 May 1914, RP, Box-2, Folder: Correspondence Apr–Sept 1914, SIO Archive.
55. Ritter to Kofoid, October 1912, RP, Ctn. 1, Folder: Outgoing Letters, 1912–14, Bancroft Archive.
56. Ibid., p. 2.
57. H. F. Osborn, 'The Present Problem of Heredity', *Atlantic Monthly*, 57 (1891), p. 363.
58. See W. E. Ritter, *The Unity of the Organism, or, the Organismal Conception of Life*, 2 vols (Boston, MA: Gorham Press, 1919) (hereafter *The Organismal Conception*), vol. 2, pp. 89–90.
59. See Ritter to Kofoid, 12 October 1912, RP, Ctn. 1, Folder: Outgoing Letters, 1912–14, Bancroft Archive.
60. Orthodox eugenics would never achieve its goals if it kept environmental forces out. To Ritter, heredity was unthinkable without development and environment: 'What eugenists of this school have failed to see, evidently, is that even were unit-factors as differentiate from one another in heredity as the extremest Mendelist conceives them to be, and that even were the germ-plasm improved up to the level of his highest hopes, his results in terms of actual human lives and social conditions would be distressingly meager. They would be so, because whether unit-factors exist independently in heredity or not, they certainly do not exist thus independently in development and function. In these ways they interact upon one another in the most vital manner, as physiology, especially of the internal secretions and nervous system, and as physiological and social psychology are rapidly and conclusively demonstrating' (Ritter, *The Organismal Conception*, vol. 2, p. 90).
61. The Ritter collection at the Bancroft Library has conserved many handwritten notes in which Ritter discusses Aristotle, Kant's *Critique of Judgment*, Blumenbach, Goethe, von Baer, Cuvier and Owen's morphology.
62. In 1931 Ritter attended the Second Congress for the History of Science held in London. He chaired a session where figures such as Haldane, Needham, Woodger and Russell

participated. On that occasion he was directly confronted with the British holist bio-philosophies. In the Bancroft Library are conserved Ritter's annotated papers that all these British scholars addressed at that congress. All treated the same topic: the relations between physical and biological sciences. Yet Ritter knew very well Russell's book on the *Interpretation of Heredity and Development.* Indeed he warmly recommended reading it in a letter addressed to von Bertalanffy in 1933 (see letter from von Bertalanffy to Ritter, 22 October 1931, RP, 71/3c, Box-6, Folder: Bertalanffy 1901, Bancroft Archive).

63. W. E. Ritter and E. W. Bailey, 'The Organismal Conception, its Place in Science and its Bearing on Philosophy', *UC-Publications in Zoology*, 31:14 (1928), pp. 307–58, on p. 307.
64. Ritter, 'Life from a Biologist's Standpoint', *Popular Science Monthly*, 75 (1909), pp. 174–90, on p. 176.
65. Ibid., p. 178.
66. Ibid., p. 180.
67. Ibid., p. 181.
68. Ritter defined vitalism as 'a walled city with the gates locked and the keys lost beyond recovery' (ibid., p. 184).
69. Ibid., p. 190.
70. Ritter, *The Organismal Conception*, vol. 1, p. 1.
71. Ritter considered himself a disciple of Aristotle.
72. Ritter, *The Organismal Conception*, vol. 1, p. 10.
73. Ibid., p. 305.
74. W. E. Ritter, 'On the Nature of Heredity and Acquired Characters, and the Question of Transmissibility of these Characters', *University Chronicle*, 3:5–6 (1903), pp. 1–24.
75. Ibid., p. 7.
76. See letter sent by Ritter to Whitman, 29 November, 1907, RP, Ctn. 1, Folder: Correspondence 1907, Bancroft Archive.
77. Ritter ascribed to Yves Delage the use of that terminology; see Delage, *L'Hérédité.*
78. W. E. Ritter, 'The Hypothesis of "Presence and Absence" in Mendelian Inheritance', *Science*, 30:768 (1909), pp. 367–8.
79. Ritter considered genetical theory a direct outcome of Weismann's germ plasm theory (Ritter, *The Organismal Conception*, vol. 1, p. 306). He believed there was a direct historical line among Weismannism, neo-Mendelism and Morganism, all tied together from an elementalist view of living phenomena.
80. Ritter, 'The Hypothesis', p. 367.
81. Ritter to E. B. Wilson, 1 June 1910, RP, Ctn. 1, Folder: Outgoing Letters, 1910–11, Bancroft Archive.
82. In other words, determiners were not the direct cause for greyness but represented only a chemical element triggering a cascade of causes/effects bringing towards the final result of greyness.
83. Ritter thought that cytoplasm played an essential role in heredity and development. In his 1919 monograph, he dedicated a massive effort to showing why cytoplasm had to be taken into account. On that debate, see Sapp, *Beyond the Gene.*
84. Ritter to E. B. Wilson, 1 June 1910, RP, Ctn. 1, Folder: Outgoing Letters, 1910–11, Bancroft Archive.
85. In the same letter, Ritter harshly criticized De Vries and his ultimate units of heredity. Not only had Ritter compared Mendelian genetics to Hegel's philosophy, he also compared De Vries's science to dogmatic theology; see ibid., p. 2.

86. Ritter to E. B. Wilson, 1 June 1910, RP, Ctn. 1, Folder: Outgoing Letters, 1910–11, Bancroft Archive.
87. See G. E. Allen, *Thomas Hunt Morgan: The Man and his Science* (Princeton, NJ: Princeton University Press, 1978).
88. See Ritter to Morgan, 15 September, 1911, RP, Ctn. 1, Folder: Outgoing Letters, 1910–11, Bancroft Archive.
89. See E. G. Conklin, 'The Mechanism of Heredity', *Science*, 27:681 (1908), pp. 89–99; and Johannsen, 'The Genotype', p. 45.
90. Ritter, *The Organismal Conception*, vol. 1, p. 312.
91. Ritter defined heredity in the following way: '"heredity" is the term applied to the universal truth that as the organism unfolds itself from the relatively minute and simple stage known as the germ into the relatively large and complex stage known as the adult, it does this in accordance with a scheme or patter characteristic of the species to which the organism belongs, so that any particular individual in the series resembles those which have gone before it ... this unfolding ... consists essentially of a great intricacy and succession of transformations' (ibid., p. 322).
92. Ritter to Kofoid, 12 October 1912, RP, Box-1, Folder: Correspondence Sept–Oct 1912, SIO Archive.
93. For this argument, see P. J. Bowler, *The Mendelian Revolution: The Emergence of Hereditarian Concepts in Modern Science and Society* (London: Continuum, 1989).
94. In 1925 Metcalf actively participated to the Scopes trial as an expert in evolutionary biology.
95. In this Metcalf totally agreed. In fact, he had responded to Ritter, arguing that 'I do not for a minute believe that determiners do it all. I suspect the whole organism reacts in everything'; see Metcalf to Ritter, 13 January 1916, RP, Ctn. 1, Folder: Outgoing Letters, 1915–16, Bancroft Archive.
96. Ritter to Metcalf, 8 February 1916, RP, Ctn. 1, Folder: Outgoing Letters, 1915–16, Bancroft Archive.
97. C. M. Child, *Senescence and Rejuvenescence* (Chicago, IL: University of Chicago Press, 1915).
98. Ritter, *The Organismal Conception*, vol. 2, p. 323.
99. See R. Amundson, *The Changing Role of the Embryo in Evolutionary Thought: Roots of Evo-Devo* (Cambridge: Cambridge University Press, 2005).
100. Bailey was a zoologist who had studied in Berkeley and became an expert in child education. She collaborated with Ritter for many years on several publications, and after Ritter's death she became his literary executor.
101. Henry Bergson too was another figure admired by Ritter. Indeed, he had a very occasional correspondence with him. At the Bancroft Library is conserved a small visiting card on which Bergson wrote: 'trouve chez lui, en rentrant à Paris, le travail que Prof. W. E. Ritter a bien voulu lui adresser [Ritter's *Life from the Biologist's Standpoint*] et le remercie de cette très intéressant étude, qui met si bien en lumière l'impossibilité d'isoler complètement quelque chose dans le domaine de la vie'; H. Bergson to Ritter, n.d., Ritter Papers, Box 6, 71\3 Folder: Correspondence and Paper, *c.* 1879–1944, Incoming Letters B, Bancroft Archive.
102. Ritter and Bailey, 'The Organismal Conception', p. 334.
103. Ritter to Scripps, 14 June 1911, Ritter Papers, Folder: Correspondence Jan–Jun 1911, SIO Archive.

104. W. E. Ritter, n.d., Ritter Papers, Ctn. 1, Folder: Biology Greater than Evolution, Bancroft Archive.
105. In Ritter's words: 'We are at last coming to see quite clearly that there can be no one all-sufficient cause of evolution; that there are, of necessity must be, a very large number of such causes. It seems pretty clear that we shall soon recognize that, taking evolution as a whole, of all sorts, part, present and future, the number of causes or "factors" must be infinite, and the business of biology on this side must be to investigate as many of these as nearly a complete fashion is possible; without, however, supposing that the task will ever be finished'; W. E. Ritter, n.d., Ritter papers, Ctn. 1, Folder: Biology Greater than Evolution, Bancroft Archive. Nevertheless, although evolution may have several causes, Ritter excluded any supernatural force behind it.
106. As Ritter argued: 'The philosophy of life that makes evolution the "master key" practically excludes from its purview the evolution of the individual. This exclusively processive biological philosophy is so dominating and many-faced, and has overflowed into so many other departments of thought, that we must run it home relentlessly. When this is done its breeding ground is found to be in exactly the same country as that of the erroneous notions about the competency of a few elements in the germ-cells to explain the adult organism which arises there from' (ibid, p. 7).
107. See Ritter and Bailey, 'The Organismal Conception', pp. 321–2.
108. Ibid., p. 322.
109. Ibid.
110. See F. B. Sumner, 'William Emerson Ritter: Naturalist and Philosopher', *Science*, new ser., 99:2574 (1944), pp. 335–8.
111. Ritter, *The Organismal Conception*, vol. 2, p. 111.
112. See Ritter to Child, 10 March 1907, Ritter Papers, Ctn. 1, Folder: Outgoing Letters 1907–9, Bancroft Archive; and H. L. Hyman, *Charles Manning Child 1869–1954* (Washington, DC: National Academy of Sciences, 1954).
113. See J. A. Ronald, 'Libbie Henrietta Hyman (1888–1969): From Developmental Mechanics to the Evolution of Animal Body Plans', *Journal of Experimental Zoology Part B: Molecular and Developmental Evolution*, 302:5 (2004), pp. 413–23.
114. Ibid., p. 79.
115. Child to Ritter, 12 July 1909, Ritter Papers, Ctn. 8, Folder: Child 1868–1954, Bancroft Archive.
116. As Haraway reminds us, drawing on Harrison, it was Boveri who first introduced the term 'gradient' in biology; see Haraway, *Crystals, Fabrics, and Fields*.
117. C. M. Child, 'Studies on the Dynamics of Morphogenesis and Inheritance in Experimental Reproduction. 1. The Axial Gradient in *Planaria dorotocephala* as a Limiting Factor in Regulation', *Journal of Experimental Zoology*, 13:1 (1911), pp. 265–320 (hereafter 'Studies 1'), on p. 268.
118. Ibid., p. 267.
119. Ibid., p. 268.
120. Child, *Senescence and Rejuvenescence*, p. v.
121. See G. Mitman and A. Fausto-Sterling, 'Whatever Happened to Planaria? C. M. Child and the Physiology of Inheritance', in A. E. Clarke and J. H. Fujimura (eds), *The Right Tools for the Job: At Work in Twentieth-Century Life Sciences* (Princeton, NJ: Princeton University Press, 1992), pp. 172–97.
122. Child, *Senescence and Rejuvenescence*, p. 19.

123. C. M. Child, *Individuality in Organisms* (Chicago, IL: University of Chicago Press, 1915), p. 4.
124. During the first three decades of the twentieth century, colloid science, as Andrew Ede showed in his book *The Rise and Decline of Colloid Science in North America, 1900–1935: The Neglected Dimension* (London: Ashgate, 2007), was a hot topic which gained progressive recognition among American scientists. Particularly in biology, it was held that the science of colloids could provide an answer to the fundamental basis of life, i.e. the protoplasm in cells possessed a kind of colloidal organization.
125. Child, *Senescence and Rejuvenescence*, p. 26.
126. Child, *Individuality*, p. 16.
127. Ibid.
128. Child, *Senescence and Rejuvenescence*, p. 6.
129. Child thoroughly denied that the origin of such an order exhibited during development could be explained with Weismann's determinants or Morgan's genes: the idea that the adult organism was a result of distinct preformed germs 'succeed[ed] merely in placing the problem beyond the reach of investigation'. Furthermore, even the most recent experimental investigations put forward by Morgan's students do 'not afford any real support to the view that the morphological characters of the adult are represented in some way by distinct entities in the germ ... and the attempts to connect particular factors with particular chromosomes or part of chromosomes are not at present, properly speaking, scientific hypotheses'; ibid., pp. 46–7.
130. Ibid., p. 56.
131. Ibid., p. 99.
132. Ibid., p. 102.
133. C. M. Child, 'Certain Dynamic Factors in Experimental Reproduction and their Significance for the Problem of Reproduction and Development', *Journal of Experimental Zoology*, 35:4 (1912), pp. 598–641, on p. 607.
134. Child to Ritter, 15 February 1927, Ritter Papers, Ctn. 8, Folder: Child 1868–1954, Bancroft Archive.
135. Child, *Individuality*, p. 59.
136. In the papers Child published between 1911 and 1913, *Planaria* was effectively the model-organism on which he based his results; see 'Studies 1', and 'Studies on the Dynamics of Morphogenesis and Inheritance in Experimental Reproduction. VI. The Nature of the Axial Gradients in *Planaria* and their Relation to Antero-Posterior Dominance, Polarity and Symmetry', *Journal of Experimental Zoology*, 37:1 (1913), pp. 108–58 (hereafter 'Studies VI'). However, when Child reiterated all this data in his 1915 books, he attempted to extend such results to other model-organisms.
137. Child, *Senescence and Rejuvenescence*, p. 199.
138. Child, 'Studies VI', p. 145.
139. Ibid., p. 109. In T. H. Morgan's book *Regeneration* (New York: Macmillan, 1901), a similar hypothesis was proposed. Regulation during regeneration was due to some specific tensions among an organism's parts.
140. Child, 'Certain Dynamic Factors', p. 604.
141. Child, *Individuality*, p. 105.
142. Child, *Senescence and Rejuvenescence*, p. 228.
143. Child, *Individuality*, p. 35.
144. 'The basis of individuality is inherited from the parents', said Child (ibid., p. 41).
145. Ibid., p. 40.

146. M. E. Sunderland, 'Regeneration: Thomas Hunt Morgan's Window into Development', *Journal of the History of Biology*, 43:2 (2010), pp. 325–61.
147. Child, 'Studies 1', p. 266.
148. Von Baer, 'Fragments related to Philosophical Zoology', p. 238.
149. Ibid., p. 239.
150. Child, *Senescence and Rejuvenescence*, p. 323.
151. Ibid.
152. Ibid., p. 347.
153. See Child, 'Certain Dynamic Factors', p. 30.
154. Child, *Senescence and Rejuvenescence*, p. 427.
155. Ibid., p. 462.
156. C. M. Child, 'The Process of Reproduction in Organisms', *Biological Bulletin*, 23:1 (1912), pp. 1–39, on p. 37.
157. As Child specified, developmental phenomena did not consist in 'a distribution of the different qualities to different regions, but simply the realization of possibilities, of capacities of the reaction system' (Child, *Individuality*, p. 202).
158. Child, 'Studies VI', p. 152.
159. Child, *Individuality*, p. 269.
160. Child, 'The Process of Reproduction', p. 35.
161. Child, *Senescence and Rejuvenescence*, pp. 462–3.
162. Child, *Individuality*, p. 205.
163. Child, 'Studies 1', p. 216.
164. Child, 'The Process of Reproduction', p. 36.
165. Child, quoted in Mitman and Fausto-Sterling, 'Whatever Happened to Planaria?', p. 15.
166. See the project for eugenics based on education in W. E. Ritter, n.d., 'A Project for Experimental Study of Development', Ritter Papers, Ctn. 11, 71/3c, Folder: Children Education Development, Bancroft Archive. See also W. E. Ritter, n.d., 'Foundation of Human Biology, Fundamentals of the Nature of Man?', Ctn. 10, 71/3, Folder: Lectures and Seminars Notes, Bancroft Archive.
167. Child, *Physiological Foundations of Behavior* (New York: Henry Holt, 1924), p. 267.
168. Ibid., p. 270.
169. The Frenchman Alfred Espinas (1844–1922), a pupil of Comte and scholar of Spencer, had published in 1878 *Des Société Animales*. The more famous Peter Kropotkin (1842–1921), a Russian thinker and anarchist, had published the celebrated *Mutual Aid: A Factor in Evolution* in 1902. In stressing the importance of cooperation in evolution, both books had an important influence on the Chicago crowd (see Mitman, *The State*). For a biographical sketch of Allee, see K. P. Schmidt, 'Warder Allee 1885–1966', in *Biographical Memoirs: National Academy of Sciences* (Washington, DC: National Academies Press, 1957), at http://www.nasonline.org/publications/biographical-memoirs/memoir-pdfs/allee-warder.pdf [accessed 13 August 2013].
170. In other words, just as in *Hydra* or *Planaria* each part or cell can reproduce the whole, in clans or tribes each member could establish new groups. To Child, integration meant a hierarchical order where the subordinated parts (governed) are tied to the dominant part (governor).
171. Child, *Physiological Foundations*, p. 288.
172. Ibid., p. 297.

173. Ibid., p. 287. And yet: 'According to the biological viewpoint, each human being is what he is because of present and past relations to environment of his protoplasm and the protoplasm from which it has arisen' (ibid., p. 286).

## Conclusion: Whatever Happened to Organismal Biologies?

1. Although, as we have seen, there were diverse meanings of mechanism. Woodger himself distinguished between two kinds of mechanist conceptions: ontological mechanism, i.e. biologists stating that organisms are actually machines, and methodological mechanism, biologists who think that whatever organisms may be, they can be investigated *as if* they were machines. These differences reflected real positions expressed by several scholars analysed by Woodger. Accordingly, there are several meanings of reductionism, materialism and holism. See Woodger, *Biological Principles*, p. 29.
2. Quoted in Kovarik, *Revolutions*, p. 78.
3. Gilbert and Sarkar, 'Embracing Complexity'.
4. Kay, *The Molecular Vision*, p. 5.
5. Ibid., p. 15.
6. Ibid., p. 92.
7. Ibid., p. 18.
8. See D. Depew and B. Weber, *Darwinism Evolving: Systems Dynamics and the Genealogy of Natural Selection* (Cambridge, MA: MIT Press, 1997); and Amundson, *The Changing Role.*
9. B. Goodwin, *How the Leopard Changed its Spots* (Princeton, NJ: Princeton University Press, 1994), p. 1.
10. See Morange, *History*.
11. See Amundson, *The Changing Role.*
12. Mitman, *The State*, p. 144.
13. The stunning similarities between Lysenko's and Child's, Ritter's or Haldane's conception of heredity and the organism appear evident while reading Lysenko's *Heredity and its Variability*: 'We, on studying heredity, ascertain the conditions of life, the conditions of development, required by the organism or by separate processes, and also the relation of the organism or separate processes of it to various environmental conditions. We thus arrive at a comprehension of the essence of heredity. The geneticists, on the other hand, do not study the essence of heredity. All they find out is how many offspring resemble the one parent with regard to a particular character and home many the other'; T. D. Lysenko, *Heredity and its Variability* (Honolulu, HI: University Press of the Pacific, 2002), p. 11. And yet: 'The development of an organism, like its growth, proceeds by means of conversion, of metabolism. The sex cells, buds or eyes from which usually entire organisms develop are, as a rule, the product of the development of the whole organism which gave rise to the particular initial basis for the new organisms. They arise and build themselves out of molecules, granules of substances – transformed many times (ibid., p. 21). As should be evident after reading this dissertation, Lysenko was not original in what he was arguing.
14. Depew and Weber, *Darwinism Evolving*, p. 395.
15. See J. Cain, 'Rethinking the Synthesis Period in Evolutionary Studies', *Journal of the History of Biology*, 42 (2009), pp. 621–48.
16. Depew and Weber, *Darwinism Evolving*, p. 395.
17. As Morange observed: 'the development of the neo-Darwinian evolutionary síntesis placed a major role in the growth of molecular biology' (Morange, *History*, p. 247).

18. See M. Laubichler, 'The Organism is Dead, Long Live the Organism', *Perspectives on Science*, 8 (2000), pp. 286–315.
19. R. C. Lewontin, S. Rose and L. J. Kamin, *Not in Our Genes: Biology, Ideology and Human Nature* (London: Penguin Books, 1984), p. 11.
20. See M. W. Elsasser, *Atom and Organism: A New Approach to Theoretical Biology* (Princeton, NJ: Princeton University Press, 1966).
21. S. Oyama, *The Ontogeny of Information: Developmental Systems and Evolution* (Durham, NC: Duke University Press, 2000), p. 39.
22. Ibid., p. 41.
23. Ibid., p. 170.
24. R. Hubbard, *Exploding the Gene Myth: How Genetic Information is Produced and Manipulated by Scientists, Physicians, Employers, Insurance Companies, Educators, and Law Enforcers* (Boston, MA: Beacon Press, 1993), p. 11.
25. E. F. Keller, *The Century of the Gene* (Cambridge, MA: Harvard University Press, 2000), p. 71.
26. Lewontin, quoted in Oyama, *The Ontogeny*, p. 11.
27. Griffiths and Gray, quoted in S. Oyama, R. G. Gray and P. E. Griffiths, *Cycles of Contingency: Developmental System and Evolution* (Cambridge, MA: MIT Press, 2003), p. 195.
28. G. B. Müller and S. Newman, *Origination of Organismal Form: The Forgotten Cause in Evolutionary Theory* (Cambridge, MA: MIT Press, 2003), p. 51.
29. C. Van der Weele, *Images of Development: Environmental Causes in Ontogeny* (Albany, NY: SUNY Press, 1999), p. 6.
30. Gilbert and Sarkar, 'Embracing Complexity', p. 1.
31. See B. K. Hall, 'Evo-Devo or Devo-Evo, Does It Matter?', *Evolution and Development*, 2:4 (2001), pp. 177–8; and G. B. Müller, 'Evo-Devo: Extending the Evolutionary Synthesis', *Nature Reviews Genetics*, 8 (2007), pp. 943–9.
32. B. K. Hall and W. M. Holson, *Keywords and Concepts in Evolutionary Developmental Biology* (Cambridge, MA: Harvard University Press, 2003), p. 61.
33. Ibid., p. 169.
34. W. Callebaut, G. Müller and S. Newman, 'The Organismic Systems Approach', in R. Sansom and R. Brandon (eds), *Integrating Evolution and Development* (Cambridge, MA: MIT Press, 2007), pp. 25–92, on p. 51.
35. See Amundson, *The Changing Role*; and M. Pigliucci and G. B. Müller, *Evolution: The Extended Synthesis* (Cambridge, MA: MIT Press, 2010).
36. See Cain, 'Rethinking the Synthesis Period'; R. Delisle, *Les philosophies du néo-darwinisme* (Paris: PUF, 2009); R. Delisle, 'What Was Really Synthesized during the Evolutionary Synthesis? A Historiographic Proposal', *Studies in History and Philosophy of Science Part C: Studies in History and Philosophy of Biological and Biomedical Sciences*, 42:1 (2011), pp. 50–9; and W. E. Reif, T. Junker and U. Hossfeld, 'The Synthetic Theory of Evolution: General Problems and the German Contribution to the Synthesis', *Theory in Biosciences*, 119:1 (2000), pp. 41–91.
37. See L. Olsson, G. S. Levit and U. Hossfeld, 'Evolutionary Developmental Biology: Its Concepts and History with a Focus on Russian and German Contributions', *Naturwissenschaften*, 97:11 (2010), pp. 951–69.
38. L. Buss, *The Evolution of Individuality* (Princeton, NJ: Princeton University Press, 1987), p. 178.

39. See M. Levin, 'Morphogenetic Fields in Embryogenesis, Regeneration, and Cancer: Non-Local Control of Complex Patterning', *Biosystems*, 109 (2012), pp. 243–61; N. Rooney, N., K. S. McCann and D. L. G. Noakes (eds), *From Energetics to Ecosystems: The Dynamics and Structure of Ecological Systems* (Dordrecht: Springer, 2007); and M. Koen-Alonso, 'A Process-Oriented Approach to the Multispecies Functional Response', in N. Rooney, K. S. McCann and D. L. G. Noakesn (eds), *From Energetics to Ecosystems: The Dynamics and Structure of Ecological Systems* (Dordrecht: Springer, 2007), pp. 1–36.
40. G. Radick, 'Other Histories, Other Biologies', *Royal Institute of Philosophy Supplement*, 56 (2005), pp. 3–4. On counterfactual history, see the articles featured in the 'Focus: Counterfactuals and the History of Science' section of the journal *ISIS*, 99:3 (2008), pp. 547–84.

# WORKS CITED

## Archival Sources

Historical Archives, Scripps Institution of Oceanography, University of California, San Diego, La Jolla, CA, USA

Bancroft Library Archive, University of California, Berkeley, CA, USA

Historical Archives, Woods Hole Oceanographic Institution, MA, USA

Special Collections, University College London, UK

Historical Archives, University of St Andrews, UK

Historical Archives, Stazione Zoologica Anton Dohrn, Naples, Italy

## Printed Sources


Abir-Am, P., 'The Biotheoretical Gathering, Trans-Disciplinary Authority and the Incipient Legitimating of Molecular Biology in the 1930s: New Perspective on the Historical Sociology of Science', *History of Science*, 25:67(1) (1987), pp. 1–70.

Allen, G. E., 'J. S. Haldane: The Development of the Idea of Control Mechanisms in Respiration', *Journal of the History of Medicine*, 22 (1967), pp. 392–412.

—, *Life Science in the Twentieth Century* (New York: Wiley, 1975).

—, *Thomas Hunt Morgan: The Man and his Science* (Princeton, NJ: Princeton University Press, 1978).

—, 'Heredity under an Embryological Paradigm: The Case of Genetics and Embryology', *Biological Bulletin*, 68 (1985), pp. 107–21.

—, 'Mechanism, Vitalism and Organicism in Late Nineteenth and Twentieth-Century Biology: The Importance of Historical Context', *Studies in History and Philosophy of Science Part C: Studies in History and Philosophy of Biological and Biomedical Sciences*, 36:2 (2005), pp. 261–83.

Amidon, K. S., 'Adolf Meyer-Abich, Holism, and the Negotiation of Theoretical Biology', *Biological Theory: Integrating Development, Cognition and Evolution*, 3:4 (2008) pp. 357–70.

Amundson, R., *The Changing Role of the Embryo in Evolutionary Thought: Roots of Evo-Devo* (Cambridge: Cambridge University Press, 2005).

Andreade, E. N., 'Obituary of E. Hatschek', *Nature*, 154:3897 (1944), pp. 46–7.

Anon., 'Obituary, William Thierry Preyer (1841–1897)', *Nature*, 56:1448 (1897), p. 296.

Anon., 'Rudolf Leuckart', *Nature*, 57:1484 (1898), p. 542.

Appell, T., *The Cuvier–Geoffrey Debate: French Biology in the Decades before Darwin* (Oxford: Oxford University Press, 1987).

B. D. J., 'Dr. John Beard', *Nature*, 114:904 (1924), p. 904.

Barlow, P. W. '"The Plant Forms Cells, Not Cells the Plant": The Origin of de Bary's Aphorism', *Annals of Biology*, 49:2 (1982), pp. 269–72.

Beckner, M., *The Biological Way of Thought* (Berkeley, CA: University of California Press, 1959).

—, 'Organismic Biology', in P. Edwards (ed.), *Encyclopedia of Philosophy* (London: Macmillan, 1969).

Beiser, F. C., *German Idealism: The Struggle against Subjectivism 1781–1801* (Cambridge, MA: Harvard University Press, 2002).

—, *The Romantic Imperative: The Concept of Early German Romanticism* (Cambridge, MA: Harvard University Press, 2004).

—, *Hegel* (London: Routledge, 2005).

—, 'Kant and Naturphilosophie', in M. Friedman and A. Nordmann (eds), *The Kant Legacy in Nineteenth-Century Science* (Cambridge, MA: MIT Press, 2006), pp. 7–26.

Ben-David, J., and T. A. Sullivan, 'Sociology of Science', *Annual Review of Sociology*, 1 (1975), pp. 203–22.

Benson, K. R., 'Why American Marine Stations?: The Teaching Argument', *American Zoologist*, 28:1 (1988), pp. 7–14.

Bernasconi, R., and T. L. Lott, *The Idea of Race* (Indianapolis, IN: Hackett Publishing Company, 2000).

Bergson, H., *L'Évolution Créatrice* (Paris: Les Presses universitaires de France, 1907).

—, *Creative Evolution* (London: Macmillan, 1922).

Berlin, I., *The Roots of Romanticism* (Princeton, NJ: Princeton University Press, 1999).

Bernard, C., *La Science Experimentale* (Paris: Bailliere, 1878).

Blackman, H., 'A Spiritual Leader? Cambridge Zoology, Mountaineering and the Death of F. M. Balfour', *Studies in History and Philosophy of Science Part C: Studies in History and Philosophy of Biological and Biomedical Sciences*, 35:1 (2003), pp. 93–117.

Blanchard, R., 'Notices biographiques: Rodolphe Leuckart', *Archives de parasitologie*, 1 (1898), pp. 185–90.

Bowler, P. J., *The Mendelian Revolution: The Emergence of Hereditarian Concepts in Modern Science and Society* (London: Continuum, 1989).

—, *Reconciling Science and Religion: The Debate in Early-Twentieth-Century Britain* (Chicago, IL: University of Chicago Press, 2001).

Brachet, A., *L'Œuf et les Facteurs de L'Ontogenèse* (Paris: G. Doin et Cie, 1917).

Brady, R. H., 'Form and Cause in Goethe's Morphology', in F. Amrine, F. J. Zucker and H. Wheele (eds), *Goethe and the Sciences: A Reappraisal* (Dordrecht: D. Reidel, 1987), pp. 257–300.

Brauckmann, S., 'The Scientific Life of Paul A. Weiss (1898–1989)', *Mendel Newsletter*, 12 (2003), pp. 2–7.

Bud, R., *The Uses of Life: A History of Biotechnology* (Cambridge: Cambridge University Press, 1993).

Buss, L., *The Evolution of Individuality* (Princeton, NJ: Princeton University Press, 1987).

Butts, R. E., 'Teleology and Scientific Method in Kant's Critique of Judgment', *Nous*, 24:1 (1990), pp. 1–16.

Byrnes, W. M., 'Ernest Everett Just, Johannes Holtfreter, and the Origin of Certain Concepts in Embryo Morphogenesis', *Molecular Reproduction and Development*, 76 (2009), pp. 912–21.

—, and W. R. Eckberg, 'Ernest Everett Just (1883–1941) – An Early Ecological Developmental Biologist', *Developmental Biology*, 296 (2006), pp. 1–11.

Cain, J., 'Woodger, Positivism and Evolutionary Synthesis', *Biology and Philosophy*, 15:4 (2000), pp. 535–51.

—, 'Rethinking the Synthesis Period in Evolutionary Studies', *Journal of the History of Biology*, 42 (2009), pp. 621–48.

Callebaut, W., G. Müller and S. Newman, 'The Organismic Systems Approach', in R. Sansom and R. Brandon (eds), *Integrating Evolution and Development* (Cambridge, MA: MIT Press, 2007), pp. 25–92.

Carter, M., *T. H. Green and the Development of Ethical Socialism* (Exeter: Imprint Academic, 2003).

Cassirer, E., *Einstein's Theory of Relativity Considered from the Epistemological Standpoint* (Chicago, IL: Open Court, 1923).

—, *The Problem of Knowledge: Philosophy, Science and History since Hegel* (New Haven, CT: Yale University Press, 1950).

Child, C. M., 'Studies on the Dynamics of Morphogenesis and Inheritance in Experimental Reproduction. 1. The Axial Gradient in *Planaria dorotocephala* as a Limiting Factor in Regulation', *Journal of Experimental Zoology*, 13:1 (1911), pp. 265–320.

—, 'Certain Dynamic Factors in Experimental Reproduction and their Significance for the Problem of Reproduction and Development', *Journal of Experimental Zoology*, 35:4 (1912), pp. 598–641.

—, 'The Process of Reproduction in Organisms', *Biological Bulletin*, 23:1 (1912), pp. 1–39.

—, 'Studies on the Dynamics of Morphogenesis and Inheritance in Experimental Reproduction. VI. The Nature of the Axial Gradients in *Planaria* and their Relation to Antero-Posterior Dominance, Polarity and Symmetry', *Journal of Experimental Zoology*, 37:1 (1913), pp. 108–58.

—, *Individuality in Organisms* (Chicago, IL: University of Chicago Press, 1915).

—, *Senescence and Rejuvenescence* (Chicago, IL: University of Chicago Press, 1915).

—, *Physiological Foundations of Behavior* (New York: Henry Holt, 1924).

—, *Biographical Memoir of Francis Bertody Sumner, 1874–1945* (Washington, DC: National Academy of Science, 1946).

Churchill, F. B., 'From Machine-Theory to Entelechy: Two Studies in Developmental Teleology', *Journal of the History of Biology*, 2:1 (1969), pp. 165–85.

—, 'William Johannsen and the Genotype Concept', *Journal of the History of Biology*, 7:1 (1974), pp. 5–30.

Cohen, D. R., 'Living Precisely in the Fin-de-Siècle Vienna', *Journal of the History of Biology*, 39:3 (2006), pp. 493–532.

Coleman, W., *Biology in the Nineteenth Century: Problems of Form, Function and Transformation* (Cambridge: Cambridge University Press, 1971).

Conklin, E. G., 'The Mechanism of Heredity', *Science*, 27:681 (1908), pp. 89–99.

Cornell, J. F., 'Faustian Phenomena: Teleology in Goethe's Interpretation of Plants and Animals', *Journal of Medicine and Philosophy*, 15 (1990), pp. 481–92.

Cunningham, A., and N. Jardine (eds), *Romanticism* and the Sciences (Cambridge: Cambridge University Press, 1990).

Cuvier, G., *Discours sur les révolutionnes de la surface du globe et sur les changements qu'elles ont produits dans le règne animal* (Paris: C. Bourgois, 1985).

Dawson, V., *Nature's Enigma: The Problem of the Polyp in the Letters of Bonnet, Trembley, and Reaumur* (Philadelphia, PA: American Philosophical Society, 1987).

Delage, Y., *L'Hérédité et les grandes problèmes de la biologie générale*, 2nd edn (Paris: Reinwald, 1903).

Delisle, R., *Les philosophies du néo-darwinisme* (Paris: PUF, 2009).

—, 'What Was Really Synthesized during the Evolutionary Synthesis? A Historiographic Proposal', *Studies in History and Philosophy of Science Part C: Studies in History and Philosophy of Biological and Biomedical Sciences*, 42:1 (2011), pp. 50–9.

Depew, D., and B. Weber, *Darwinism Evolving: Systems Dynamics and the Genealogy of Natural Selection* (Cambridge, MA: MIT Press, 1997).

Desmond, A., *The Politics of Evolution: Morphology, Medicine, and Reform in Radical London* (Chicago, IL: University of Chicago Press, 1989).

Dewey, J., *The Influence of Darwin on Philosophy and Other Essays* (New York: Henry Holt, 1910).

Dinsmore, C. E., *History of Regeneration Research: Milestones in the Evolution of a Science* (Cambridge: Cambridge University Press, 1991).

Drack, M., W. Apfalter and D. Pouvreau, 'On the Making of a System Theory of Life, Paul A. Weiss and Ludwig von Bertalanffy's Conceptual Connections', *Quarterly Review of Biology*, 82:4 (2007), pp. 349–73.

Dronamraju, K. R., *If I Am to be Remembered: The Life and Work of Julian Huxley with Selected Correspondence* (Singapore: World Scientific, 1993).

Ede, A., *The Rise and Decline of Colloid Science in North America, 1900–1935: The Neglected Dimension* (London: Ashgate, 2007).

Edwards, C. L., 'The Vienna Institution for Experimental Biology', *Popular Science Monthly*, 78:37 (1911), pp. 584–601.

Enriques, F., 'Motivi dell Filosofia di Eugenio Rignano', *Scientia*, 24:47 (1930), pp. 337–84.

Elsasser, M. W., *Atom and Organism: A New Approach to Theoretical Biology* (Princeton, NJ: Princeton University Press, 1966).

Esposito, J. L., *Schelling's Idealism and Philosophy of Nature* (Lewisburg, PA: Bucknell University Press, 1977).

Farley, J., *Gametes and Spores, Ideas about Sexual Reproduction: 1750–1914* (Baltimore, MD: Johns Hopkins University Press, 1982).

Ferrini, C., 'The Transition to Organics: Hegel's Idea of Life', in S. Houlgate and M. Baur (eds), *A Companion to Hegel* (Oxford: Wiley-Blackwell, 2011), pp. 203–24.

Fink, K. J., *Goethe's History of Science* (Cambridge: Cambridge University Press, 1991).

Foster, M., and F. M. Balfour, *The Elements of Embryology* (London: Macmillan, 1874).

Friedman, M., and A. Nordmann (eds), *The Kantian Legacy in Nineteenth-Century Science* (Cambridge, MA: MIT Press, 2006).

Gayon, J., 'History of the Concept of Allometry', *American Zoologist*, 40:5 (2000), pp. 748–58.

Garstand, W., 'The Theory of Recapitulation: A Critical Re-statement of the Biogenetic Law', *Linnean Journal-Zoology*, 35:232 (1922), pp. 81–101.

Gasking, E., *Investigations into Generation* (Baltimore, MD: Johns Hopkins University Press, 1968).

Geison, G. L., *Michael Foster and the Cambridge School of Physiology: The Scientific Enterprise in Late Victorian Society* (Princeton, NJ: Princeton University Press, 1978).

Gerabek, W. E., *Friedrich Wilhelm Joseph Schelling und die Medizin der Romantik: Studien zu Schellings Würzburger Periode*, Europäische Hochschulschriften, ser. 7, sect. B, vol. 7 (Frankfurt: Peter Lang, 1995).

Ghiselin, M. T., and C. Groeben, 'A Bioeconomic Perspective on the Organization of the Naples Zoological Station', in M. T. Ghiselin and A. E. Leviton (eds), *Cultures and Institutions of Natural History: Essays in the History and Philosophy of Science* (San Francisco, CA: California Academy of Sciences, 2000), pp. 273–85.

Gilbert, F. S., 'Cellular Politics: Ernest Everett Just, Richard B. Goldschmidt, and the Attempt to Reconcile Embryology and Genetics', in R. Rainger, K. R. Benson and J. Maienschein (eds), *The American Development of Biology* (Philadelphia, PA: University of Pennsylvania Press, 1991), pp. 311–46.

—, 'Bearing Crosses: The Historiography of Genetics and Embryology', *American Journal of Medical Genetics*, 76:2 (1998), pp. 168–82.

—, and S. Sarkar, 'Embracing Complexity: Organicism for the 21st Century', *Developmental Dynamics*, 219:1 (2000), pp. 1–9.

Ginsborg, H., 'Kant on Understanding Organisms as Natural Purposes', in E. Watkins (ed.), *Kant and the Sciences* (Oxford: Oxford University Press, 2001), pp. 231–59.

Gliboff, S., 'The Case of Paul Kammerer: Evolution and Experimentation in the Early 20th Century', *Journal of the History of Biology*, 39:3 (2006), pp. 525–63.

Goethe, J. W. von, *Faust* (New York: Bantam Classics, 1988).

Goodfield, G. J., *The Growth of Scientific Physiology* (London: Hutchinson, 1960).

Goodman, M., *Suffer and Survive* (London: Simon & Schuster, 2007).

Goodwin, B., *How the Leopard Changed its Spots* (Princeton, NJ: Princeton University Press, 1994).

Gould, S. J., 'D'Arcy Thompson and the Science of Form', *New Literary Hist*ory, 2:2 (1971), pp. 229–58.

—, *Ontogeny and Phylogeny* (Cambridge, MA: Harvard University Press, 1977).

—, 'Thwarted Genius', *New York Review of Books*, 30:18 (24 November 1983), at http://www.nybooks.com/articles/archives/1983/nov/24/thwarted-genius/ [accessed 13 August 2013].

—, 'Just in the Middle: A Solution to the Mechanist-Vitalist Controversy', in *The Flamingo's Smile: Reflections in Natural History* (New York: W. W. Norton, 1987), pp. 337–91.

—, *The Structure of Evolutionary Theory* (Cambridge, MA: Harvard University Press, 2002).

Graham, M., 'E. S. Russell 1887–1954', *ICES Journal of Marine Science*, 20:2 (1954), pp. 135–9.

Grant, I. H., *Philosophies of Nature after Schelling* (London: Bloomsbury Academic, 2008).

Gregg, J. R., and F. T. C. Harris, *Form and Strategy in Science: Studies Dedicated to Joseph Henry Woodger on the Occasion of his Seventieth Birthday* (Dordrecht: D. Reidel, 1964).

Gregory, F., 'Kant's Influence on Natural Scientists in the German Romantic Period', in R. P. W. Visser (ed.), *New Trends in the History of Science* (Amsterdam: Rodopi, 1989), pp. 53–66.

Guyer, P., 'Organisms and the Unity of Science', in E. Watkins (ed.), *Kant and the Sciences* (Oxford: Oxford University Press, 2001), pp. 259–82.

Hahn, S. S., *Contradiction in Motion: Hegel's Organic Concept of Life and Value* (Ithaca, NY: Cornell University Press, 2007).

Haines, C., 'Russell, Edward Stuart (1887–1954)', *Oxford Dictionary of National Biography* (Oxford: Oxford University Press, 2004).

Haldane, J. S., 'Life and Mechanism', *Mind*, 9:33 (1884), pp. 27–47.

—, *Mechanism, Life and Personality: An Examination of the Mechanistic Theory of Life and Mind* (London: Murray, 1913).

—, *Organism and Environment as Illustrated by the Physiology of Breathing* (New Haven, CT: Yale University Press, 1917).

—, *The New Physiology and Other Addresses* (London: C. Griffin, 1919).

—, *The Philosophical Basis of Biology* (London: Hodder and Stoughton, 1931).

—, *Materialism* (London: Hodder and Stoughton, 1932).

—, *The Philosophy of a Biologist* (Oxford: Clarendon Press, 1935).

—, and R. S. Haldane, 'The Relation of Philosophy to Science', in A. Seth and R. S. Haldane (eds), *Essays in Philosophical Criticism* (New York: Burt Franklin, 1883), pp. 41–66.

—, D. W. Thompson, P. C. Mitchell and L. T. Hobhouse, 'Symposium: Are Physical, Biological and Psychological Categories Irreducible?', *Proceedings of the Aristotelian Society: Supplementary Volumes*, new ser., 18 (1918), pp. 419–78.

Hall, B. K., 'Balfour, Garstang and de Beer: The First Century of Evolutionary Embryology', *American Zoologist*, 40:5 (2000), pp. 718–28.

—, 'Evo-Devo or Devo-Evo, Does It Matter?', *Evolution and Development*, 2:4 (2001), pp. 177–8.

—, 'Francis Maitland Balfour (1851–1882): A Founder of Evolutionary Embryology', *Journal of Experimental Zoology*, 299 (2003), pp. 3–8.

—, and W. M. Olson, *Keywords and Concepts in Evolutionary Developmental Biology* (Cambridge, MA: Harvard University Press, 2003).

Hamburger, V., *The Heritage of Experimental Embryology, Hans Spemann and the Organizer* (Oxford: Oxford University Press, 1988).

Hangstrom, W. O., *The Scientific Community* (Carbondale, IL: Southern Illinois University Press, 1975).

Haraway, D., *Crystals, Fabrics, and Fields* (Baltimore, MD: Johns Hopkins University Press, 1976).

Harrington, A., *Reenchanted Science: Holism in German Culture from Wilhelm II to Hitler* (Princeton, NJ: Princeton University Press, 1996).

Harwood, J., *Styles of Scientific Thought: The German Genetics Community, 1900–33* (Chicago, IL: University of Chicago Press, 2003).

Hatschek, E., 'A Study of the Forms Assumed by Drops and Vortices of a Gelatinising Liquid in Various Coagulating Solutions', *Proceedings of the Royal Society of London. Series A, Containing Papers of a Mathematical and Physical Character (1905–1934)*, 95:669 (1918), pp. 303–16.

Helferich, G., *Humboldt's Cosmos: Alexander von Humboldt and the Latin American Journey that Changed the Way We See the World*, Kindle edn (London: Penguin, 2011).

Henderson, L. J., *The Fitness of the Environment* (New York: Macmillan, 1913).

Heuser, M. L., 'Schelling's Concept of Self-Organization', in R. Friedrich and A. Wunderlin (eds), *Evolution of Dynamical Structures in Complex Systems* (Berlin: Springer, 1992), pp. 395–415.

Hine, R. V., *California's Utopian Colonies* (San Marino, CA: Huntington Library, 1953).

Hyman, H. L., *Charles Manning Child 1869–1954* (Washington, DC: National Academy of Sciences, 1954).

Hofstadter, R., *Anti-Intellectualism in American Life* (New York: Random House, 1963).

Holland, J., *German Romanticism and Science: The Procreative Poetics of Goethe, Novalis and Ritter* (London: Routledge, 2009).

Hubbard, R., *Exploding the Gene Myth: How Genetic Information is Produced and Manipulated by Scientists, Physicians, Employers, Insurance Companies, Educators, and Law Enforcers* (Boston, MA: Beacon Press, 1993).

Huneman, P. (ed.), *Understanding Purpose: Kant and the Philosophy of Biology* (Rochester, NY: University of Rochester Press, 2007).

Huxley, J. S., *Problems of Relative Growth* (London: Methuen, 1932).

—, *Evolution: The Modern Synthesis* (London: Allen & Unwin, 1954).

—, and G. Teissier, 'Terminology of Relative Growth', *Nature*, 137 (1936), pp. 780–1.

Jardine, N., *The Scenes of Inquiry: On the Reality of Questions in the Sciences* (Oxford: Oxford University Press, 1991).

Johannsen, W., *Elemente der exakten Erblichkeitslehere* (Jena: Gustav Fisher, 1909).

—, 'The Genotype Conception of Heredity', *American Naturalist*, 45 (1911), pp. 129–59.

—, 'Some Remarks about Units in Heredity', *Hereditas*, 4 (1923), pp. 133–41.

Just, E. E., 'On the Origins of Mutations', *American Naturalist*, 66 (1932), pp. 61–74.

—, 'Cortical Cytoplasm and Evolution', *American Naturalist*, 66 (1933), pp. 22–9.

—, 'A Single Theory for the Physiology of Development and Genetics', *American Naturalist*, 70 (1936), pp. 267–312.

—, 'Phenomena of Embryogenesis and their Significance for a Theory of Development and Heredity', *American Naturalist*, 71 (1937), pp. 97–112.

—, *The Biology of the Cell Surface* (London: Technical Press, 1939).

—, 'Unsolved Problems of General Biology', *Physiological Zoology*, 13:2 (1940), pp. 123–42.

Kant, I., *Kritik of Judgment*, trans. and intro. J. H. Bernard, rev. 2nd edn (London: Macmillan, 1914).

Kay, L., *The Molecular Vision of Life: Caltech, the Rockefeller Foundation, and the Rise of the New Biology* (Oxford: Oxford University Press, 1993).

Keller, E. F., *The Century of the Gene* (Cambridge, MA: Harvard University Press, 2000).

Kepner, W. A., and W. C. Whitlock, 'Food-Reactions of Pelomyxa carolinensis', *Journal of Experimental Zoology*, 24 (1917), pp. 381–404.

Kimball, G. K., 'D'Arcy Thompson: His Conception of the Living Body', *Philosophy of Science*, 20:2 (1953), pp. 139–48.

Kitchen, P., *A Most Unsettling Person: An Introduction to the Ideas and Life of Patrick Geddes* (London: Gollancz, 1975).

Koen-Alonso, M., 'A Process-Oriented Approach to the Multispecies Functional Response', in N. Rooney, K. S. McCann and D. L. G. Noakesn (eds), *From Energetics to Ecosystems: The Dynamics and Structure of Ecological Systems* (Dordrecht: Springer, 2007), pp. 1–36.

Koestler, A., *The Case of the Midwife Toad* (London: Hutchinson, 1971).

Kohler, R. E., *Landscapes and Labscapes: Exploring the Lab-Field Border in Biology* (Chicago, IL: University of Chicago Press, 2002).

Kovarik, B., *Revolutions in Communication: Media History from Gutenberg to the Digital Age* (New York: Continuum, 2001).

Kuklick, B., *Josiah Royce: An Intellectual Biography* (Indianapolis, IN: Hackett Publishing Company, 1985).

—, *A History of Philosophy in America: 1720–2000* (Oxford: Oxford University Press, 2001).

Lamb, D., *Hegel: From Foundation to System* (The Hague: Martinus Nijhoff, 1981).

Laubichler, M., 'The Organism is Dead, Long Live the Organism', *Perspectives on Science*, 8 (2000), pp. 286–315.

—, and J. Maienschein, *From Embryology to Evo-Devo: A History of Developmental Evolution* (Cambridge, MA: MIT Press, 2007).

Le Gros Clark, W., and P. Medawar, *Essays on Growth and Form: Presented to D'Arcy Wentworth Thompson* (Oxford: Clarendon Press, 1945).

Lenoir, T., *The Strategy of Life: Teleology and Mechanics in Nineteenth-Century German Biology* (Chicago, IL: University of Chicago Press, 1982).

—, 'The Eternal Laws of Form', in F. Amrine, F. J. Zucker and H. Wheeler (eds), *Goethe and the Sciences: A Reappraisal* (Dordrecht: D. Reidel, 1987), pp. 17–28.

Levin, M., 'Morphogenetic Fields in Embryogenesis, Regeneration, and Cancer: Non-Local Control of Complex Patterning', *Biosystems*, 109 (2012), pp. 243–61.

Lewontin, R. C., and R. Levins, *The Dialectical Biologist* (Cambridge, MA: Harvard University Press, 1985).

—, S. Rose and L. J. Kamin, *Not in Our Genes: Biology, Ideology and Human Nature* (London: Penguin Books, 1984).

Lynch, J. M., *Vestiges and the Debate before Darwin* (London: Thoemmes Press, 2000).

Lickliter, R., and T. D. Berry, 'The Phylogeny Fallacy: Developmental Psychology's Misapplication of Evolutionary Theory', *Developmental Review*, 10 (1990), pp. 322–38.

Lightman, B., *The Dictionary of Nineteenth-Century British Scientists* (London: Thoemmes Continuum, 2004).

Lillie, F. R., 'The Organization of the Egg in *Unio*, Based on a Study on its Maturation, Fertilization, and Cleavage', *Journal of Morphology*, 17 (1901), pp. 277–92.

—, 'Observations and Experiments Concerning the Elementary Phenomena of Embryonic Development in *Chaetopterus*', *Journal of Experimental Zoology*, 3 (1906), pp. 153–268.

—, 'Studies on Fertilization: VI. The Mechanism of Fertilization in *Arbacia*', *Journal of Experimental Zoology*, 16 (1914), pp. 523–90.

—, 'The Theory of Individual Development', *Popular Science Monthly*, 75:14 (1914), pp. 239–52.

—, 'The Gene and the Ontogenetic Process', *Science*, 66:1712 (1927), pp. 361–8.

—, 'Embryonic Segregation and its Role in the Life History', *Development Genes and Evolution*, 118:1 (1929), pp. 499–533.

—, 'The Physiology of Feather Pattern', *Wilson Bulletin*, 44:4 (1932), pp. 193–211.

—, 'Physiology of Development of the Feather. II. General Principles of Development with Special Reference to the After-Feather', *Physiological Zoology*, 11:4 (1938), pp. 434–50.

—, 'Zoological Sciences in the Future', *Science*, 88:2273 (1938), pp. 65–72.

—, 'Physiology of Development of the Feather. V. Experimental Morphogenesis', *Physiological Zoology*, 14:2 (1941), pp. 103–35.

—, 'On the Development of Feathers', *Biological Reviews*, 17:3 (1942), pp. 247–66.

—, *The Woods Hole Marine Biological Laboratory* (Chicago, IL: University of Chicago Press, 1944).

—, and R. Juhn, 'The Physiology of Development of Feathers. I. Growth-Rate and Pattern in the Individual Feather', *Physiological Zoology*, 5:1 (1932), pp. 124–84.

Linguerri, S., *La Grande Festa della Scienza. Eugenio Rignano e Federico Enriques. Lettere* (Milan: Franco Angeli, 2005).

Litwin, T., *The Harriman Alaska Expedition Retraced: A Century of Change, 1899–2001* (New Brunswick, NJ: Rutgers University Press, 2005).

Lovejoy, A., 'Kant and Evolution. II', *Popular Science Monthly*, 78 (1911), pp. 36–51.

—, 'On the Discrimination of Romanticisms', *Publications of the Modern Language Association of America*, 39 (1924), pp. 229–53.

Lurie, E., *Louis Agassiz: A Life in Science* (Chicago, IL: University of Chicago Press, 1960).

Lysenko, T. D., *Heredity and its Variability* (Honolulu, HI: University Press of the Pacific, 2002).

MacFarland, J., *Kant's Concept of Teleology* (Edinburgh: Edinburgh University Press, 1970).

Maienschein, J., 'Agassiz, Hyatt, Whitman and the Birth of Marine Biological Laboratory', *Biological Bulletin*, 168 (1985), pp. 26–34.

—, 'H. N. Martin and W. K. Brooks: Exemplars for American Biology?', *American Zoologist*, 27:3 (1987), pp. 773–83.

—, 'The Origins of Entwicklungsmechanik', in S. F. Gilbert (ed.), *A Conceptual History of Modern Embryology* (New York: Plenum Press, 1991), pp. 43–59.

—, *Transforming Traditions in American Biology, 1880–1915* (Baltimore, MD: Johns Hopkins University Press, 1991).

—, 'Whitman at Chicago: Establishing a Chicago Style of Biology?', in R. Rainger, K. R. Benson and J. Maienschein (eds), *The American Development of Biology* (Philadelphia, PA: University of Pennsylvania Press, 1991), pp. 151–80.

Manning, K. R., *Black Apollo of Science: The Life of Ernest Everett Just* (Oxford: Oxford University Press, 1983).

Mayr, E., *Toward a New Philosophy of Biology: Observations of an Evolutionist* (Cambridge, MA: Harvard University Press, 1988).

—, 'The Autonomy of Biology: The Position of Biology among the Sciences', *Quarterly Review of Biology*, 71:1 (1996), pp. 97–106.

McDowall, J. S., 'Sir Michael Foster', *Post-Graduate Medical Journal*, 12 (1936), pp. 78–9.

McLaughlin, P., *Kant's Critique of Teleology in Biological Explanation, Antinomy and Teleology* (New York: Edwin Mellen Press, 1990).

—, 'Kant on Heredity and Adaptation', in S. Müller-Wille and H. J. Rheinberger (eds), *Heredity Produced: At the Crossroads of Biology, Politics, and Culture, 1500–1870* (Cambridge, MA: MIT Press, 2007), pp. 277–91.

McWilliams, C., *Southern California: An Island on the Land* (Layton, UT: Gibbs Smith, 1973).

Meyer-Abich, A., *Ideen und Ideale der biologischen Erkenntnis: Beitrage zur Theorie und Geschichte der biologischen Ideologien* (Leipzig: J. A. Barth, 1934).

—, *Biologie der Goethezeit* (Stuttgart: Hippokrates Verlag, 1949).

Meyer-Abich, K. M., 'Self-Knowledge, Freedom and Irony: The Language of Nature in Goethe', in F. Amrine, F. J. Zucker and H. Wheeler (eds), *Goethe and the Sciences: A Reappraisal* (Dordrecht: D. Reidel, 1987), pp. 351–71.

Mitman, G., *The State of Nature: Ecology, Community, and American Social Thought, 1900–1950* (Chicago, IL: University of Chicago Press, 1992).

—, and A. Fausto-Sterling, 'Whatever Happened to Planaria? C. M. Child and the Physiology of Inheritance', in A. E. Clarke and J. H. Fujimura (eds), *The Right Tools for the Job: At Work in Twentieth-Century Life Sciences* (Princeton, NJ: Princeton University Press, 1992), pp. 172–97.

Montiel, L., 'Más allá de El Nacimiento de la clinica. La comprensión de la Anatomia general de Bichat desde la Naturphilosophie de Schelling', in O. Market and J. R. de Rosales (eds), *El inicio del Idealismo Alemán* (Madrid: Editorial Complutense, 1996), pp. 315–25.

—, 'Filosofía de la ciencia médica en el romanticismo alemán. La propuesta de Ignaz Döllinger (1770–1841) para el estudio de la fisiología', *Medicina y Historia*, 70 (1997), pp. 1–15.

Morange, M., *A History of Molecular Biology* (Cambridge, MA: Harvard University Press, 2000).

Morgan, T. H., *Regeneration* (New York: Macmillan, 1901).

—, 'Genetics and Physiology of Development', *American Naturalist*, 60:671 (1926), pp. 489–515.

—, *The Theory of the Gene* (New Haven, CT: Yale University Press, 1926).

Morse, E. S., *Biographical Memoir of Charles Otis Whitman, 1842–1910* (Washington, DC: National Academy of Sciences, 1912).

—, *Biographical Memoir of Charles Sedgwick Minot 1852–1914* (Washington, DC: National Academy of Sciences, 1920).

Moscoso, J., 'Experimentos de Regeneración Animal: 1686–1765. Come defender la Preexistencia?', *Dynamis*, 15 (1995), pp. 341–73.

Müller, G. B., 'Evo-Devo: Extending the Evolutionary Synthesis', *Nature Reviews Genetics*, 8 (2007), pp. 943–9.

—, and S. Newman, *Origination of Organismal Form: The Forgotten Cause in Evolutionary Theory* (Cambridge, MA: MIT Press, 2003).

Müller, G. H., 'Wechselwirkung in the Life and Other Sciences: A Word, New Claims, and a Concept around 1800 ... and Much Later', in S. Bossi and M. Poggi (eds), *Romanticism in Science: Science in Europe, 1790–1840* (Dordrecht: Kluwer Academic, 1994), pp. 1–14.

Müller, J., *Handbuch der Physiologie des Menschen: für Vorlesungen*, Vol. 1 (Coblenz: Verlag von J. Holscher, 1837).

Müller, H., *The Fertilization of Flowers*, ed. and trans. D. W. Thompson, with a preface by C. Darwin (London: Macmillan, 1883).

Müller, I., 'The Impact of the Zoological Station in Naples on Developmental Physiology', *International Journal of Developmental Biology*, 40 (1996), pp. 103–11.

Mulnard, J. G., 'The Brussels School of Embryology', *International Journal of Developmental Biology*, 36 (1992), pp. 17–24.

Needham, J., *Man a Machine: In Answer to a Romantical and Unscientific Treatise Written by Sig. Eugenio Rignano & Entitled 'Man Not a Machine'* (London: Kegan Paul, Trench, Trübner and Co., 1928).

—, 'Organicism in Biology', *Journal of Philosophical Studies*, 3:9 (1928), pp. 29–40.

Newmann, H. H., 'History of the Department of Zoology in the University of Chicago', *Bios*, 19:4 (1948), pp. 215–39.

Ninham, N. B., and P. Lo Nostro, *Molecular Forces and Self-Assembly* (Cambridge: Cambridge University Press, 2010).

Nicolson M., 'Alexander von Humboldt and the Geography of Vegetation', in A. Cunningham and N. Jardine (eds), *Romanticism and the Sciences* (Cambridge: Cambridge University Press, 1990), pp. 169–88.

Nyhart, L. K., *Biology Takes Form: Animal Morphology and the German Universities, 1800–1900* (Chicago, IL: University of Chicago Press, 1995).

—, and S. Lidgard, 'Individuals at the Center of Biology: Rudolf Leuckart's *Polymorphismus* and the Ongoing Narrative of Parts and Wholes. With an Annotated Translation', *Journal of the History of Biology*, 44:3 (2011), pp. 373–443.

Norcross, B. L., 'Beyond Serenity: Boom and Bust in Prince William Sound', in T. Litwin (ed.), *The 1899 Harriman Alaska: A Century of Change* (New Brunswick, NJ: Rutgers University Press, 2005), pp. 115–24.

Nordenskiöld, E., *The History of Biology* (New York: Knopf, 1935).

Olsson, L., G. S. Levit and U. Hossfeld, 'Evolutionary Developmental Biology: Its Concepts and History with a Focus on Russian and German Contributions', *Naturwissenschaften*, 97:11 (2010), pp. 951–69.

Oppenheimer, J., *Autobiography of Karl Ernst von Baer* (New York: Science History Publications, 1986).

Osborn, H. F., 'The Present Problem of Heredity', *Atlantic Monthly*, 57 (1891), p. 363.

Ospovat, D., 'The Influence of Karl Ernst von Baer's Embryology, 1828–1859: A Reappraisal in Light of Richard Owen's and William B. Carpenter's Paleontological Application of von Baer's Law', *Journal of the History of Biology*, 9:1 (1976), pp. 1–28.

Outram, D., 'Uncertain Legislator: Georges Cuvier's Laws of Nature in their Intellectual Context', *Journal of the History of Biology*, 19:3 (1986), pp. 323–68.

Overton, J., 'Paul Alfred Weiss 1898–1989', in *Biographical Memoirs: National Academy of Sciences* (Washington, DC: National Academies Press, 1997), at http://www.nap.edu/readingroom.php?book=biomems&page=pweiss.html [accessed 13 August 2013].

Owen, R., *The Hunterian Lectures in Comparative Anatomy, May and June 1837*, intro. and commentary by P. R. Sloan (Chicago, IL: University of Chicago Press, 1992).

Oyama, S., *The Ontogeny of Information: Developmental Systems and Evolution* (Durham, NC: Duke University Press, 2000).

—, R. G. Gray and P. E. Griffiths, *Cycles of Contingency: Developmental System and Evolution* (Cambridge, MA: MIT Press, 2003).

Pagel, J., *Biographical Encyclopedia of Outstanding Physicians of the Nineteenth Century*, Vol. 1550–1 (Vienna: Verlag Urban and Schwarzenberg, 1901).

Parascandola, J., 'Organismic and Holistic Concepts in the Thought of L. J. Henderson', *Journal of the History of Biology*, 4:1 (1971), pp. 63–113.

Pauly, P. J., 'The Appearance of Academic Biology in Late Nineteenth-Century America', *Journal of the History of Biology*, 17:3 (1984), pp. 369–97.

—, *Biologists and the Promise of American Life: From Meriwether Lewis to Alfred Kinsey* (Princeton, NJ: Princeton University Press, 2000).

Pasteels, J., 'Notice sur Albert Dalcq', *Annuaire de l'Academie Royale de Belgique*, 129 (1974), pp. 155–66.

Persell, S. M., *Neo-Lamarckism and the Evolution Controversy in France, 1870–1920* (New York: Edwin Mellen Press, 1999).

Peters, P., 'Frank Lillie: The Cultivator of an Embryonic Woods Hole Scientific Community', *Cape Cod Times*, 6 June 1999, at http://www.capecodonline.com/apps/pbcs.dll/article?AID=/19990606/NEWS01/306069977&cid=sitesearch&template=printart [accessed 13 August 2013].

Pigliucci, M., and G. B. Müller, *Evolution: The Extended Synthesis* (Cambridge, MA: MIT Press, 2010).

Planck, M., 'Letter', *Nature*, 127:3207 (1931), p. 612.

Poggi, S., and M. Bossi (eds), *Romanticism in Science: Science in Europe, 1790–1840* (Dordrecht: Kluwer Academic, 1994).

Pouvreau, D., *Une Biographie Non Officielle de Ludwig von Bertalanffy (1901–1972)* (Vienna: Bertalanffy Center for the Study of Systems Science, 2006), at http://www.bertalanffy.org/media/pdf/pdf29.pdf [accessed 13 August 2013].

Preyer, W. T., *Eléments de physiologie générale* (Paris: Alcan, 1884).

Provine, W. B., 'Francis B. Sumner and the Evolutionary Synthesis', *Studies in History of Biology*, 3 (1979), pp. 211–40.

Przibram, H., 'Regeneration und Transplantation im Tierreich', *Kultur der Gegenwart, Band Allgemeine Biologie*, (1913), pp. 347–77.

—, *Aufbau mathematischer Biologie*, Abhandlungen zur theoretischen Biologie, Vol. 18 (Berlin: Gebrüder Borntraeger, 1923).

Radick, G., 'Other Histories, Other Biologies', *Royal Institute of Philosophy Supplement*, 56 (2005), pp. 3–4.

Rainger, R., K. R. Benson and J. Maienschein (eds), *The American Development of Biology* (New Brunswick, NJ: Rutgers University Press, 1991).

Raitt, H., and B. Moulton, *Scripps Institution of Oceanography: First Fifty Years* (Los Angeles, CA: Ward Ritchie Press, 1966).

Reddick, J., 'The Shattered Whole: Georg Buchner and Naturphilosophie', in A. Cunningham and N. Jardine (eds), *Romanticism and the Sciences* (Cambridge: Cambridge University Press, 1990), pp. 322–40.

Rehbock, P., *The Philosophical Naturalists: Themes in Early Nineteenth-Century British Biology* (Madison, WI: University of Wisconsin Press, 1983).

Reif, W. E., T. Junker and U. Hossfeld, 'The Synthetic Theory of Evolution: General Problems and the German Contribution to the Synthesis', *Theory in Biosciences*, 119:1 (2000), pp. 41–91.

Reynolds, A., 'The Cell's Journey: From Metaphorical to Literal Factory', *Endeavour*, 31:2 (2007), pp. 65–70.

Richards, R. J., *The Meaning of Evolution: The Morphological Construction and Ideological Reconstruction of Darwin's Theory* (Chicago, IL: University of Chicago Press, 1992).

—, *The Romantic Conception of Life: Science and Philosophy in the Age of Goethe* (Chicago, IL: University of Chicago Press, 2002).

—, 'Goethe's Use of Kant in the Erotics of Nature', in P. Huneman (ed.), *Understanding Purpose: Kant and the Philosophy of Biology* (Rochester, NY: University of Rochester Press, 2007), pp. 137–48.

—, *The Tragic Sense of Life: Ernst Haeckel and the Struggle over Evolutionary Thought* (Chicago, IL: University of Chicago Press, 2008).

Richmond, M., 'A Lab of One's Own: The Balfour Biological Laboratory for Women at Cambridge University, 1884–1914', *ISIS*, 88:3 (1997), pp. 422–55.

—, 'Sedgwick, Adam (1854–1913)', *Oxford Dictionary of National Biography* (Oxford: Oxford University Press, 2004).

Ridley, M., 'Embryology and Classical Zoology in Great Britain', in T. J. Horder, J. A. Witkowski and C. C. Wylie (eds), *A History of Embryology* (Cambridge: Cambridge University Press, 1986), pp. 35–68.

Rignano, E., *Sur la Transmissibilité des Caractères Acquis: Hypothèse d'une Centro-épigénèse* (Paris: Félix Alcan, 1906).

—, *Man Not a Machine: A Study of the Finalistic Aspects of Life*, with a foreword by H. Driesch (London: Kegan Paul, Trench, Trübner and Co., 1926).

—, *Das Gedachtnis als Grundlage des Lebendigen*; *mit einer einleitung von Ludwig von Bertalanff* (Vienna: Wilhelm Braumuller, 1931).

Risse, G. B., 'Kant, Schelling, and the Early Search for a Philosophical "Science" of Medicine in Germany', *Journal of the History of Medicine*, 27:2 (1972), pp. 145–58.

Ritter, W. E., 'On the Nature of Heredity and Acquired Characters, and the Question of Transmissibility of these Characters', *University Chronicle*, 3:5–6 (1903), pp. 1–24.

—, 'Organization in Scientific Research', *Popular Science Monthly*, 67 (1904), pp. 49–53.

—, 'Life from a Biologist's Standpoint', *Popular Science Monthly*, 75 (1909), pp. 174–90.

—, 'The Hypothesis of "Presence and Absence" in Mendelian Inheritance', *Science*, 30:768 (1909), pp. 367–8.

—, *The Marine Biological Station of San Diego, its History, Present Conditions, Achievements, and Aims* (Berkeley, CA: University of California Press, 1912).

—, *The Unity of the Organism, or, the Organismal Conception of Life*, 2 vols (Boston, MA: Gorham Press, 1919).

—, and E. W. Bailey, 'The Organismal Conception, its Place in Science and its Bearing on Philosophy', *UC-Publications in Zoology*, 31:14 (1928), pp. 307–58.

—, L. L. Whyte, L. Hogben, J. S. Haldane, H. J. Woodger, E. S. Russell, A. Yoffe and J. Needham, 'Historical and Contemporary Relationships of the Physical and Biological Sciences', Second International Congress on the History of Science (Third Session), *Archeion*, 14:4 (1933), pp. 497–502.

Ritterbush, P. C., *The Art of Organic Form* (Washington, DC: Smithsonian Institution Press, 1968).

Robins, E. P., *Some Problems of Lotze's Theory of Knowledge* (New York : Macmillan, 1900).

Robbins, P., *The British Hegelians 1875–1925* (London: Garland, 1982).

Roger, J., *Les Sciences de la Vie dans la Pensée Français du XVIII Siècle* (Paris: Armand Colin, 1963).

Roll-Hansen, N., 'E. S. Russell and J. H. Woodger: The Failure of Two Twentieth-Century Opponents of Mechanistic Biology', *Journal of the History of Biology*, 17:3 (1984), pp. 399–428.

Ronald, J. A., 'Libbie Henrietta Hyman (1888–1969): From Developmental Mechanics to the Evolution of Animal Body Plans', *Journal of Experimental Zoology Part B: Molecular and Developmental Evolution*, 302:5 (2004), pp. 413–23.

Rooney, N., K. S. McCann and D. L. G. Noakes (eds), *From Energetics to Ecosystems: The Dynamics and Structure of Ecological Systems* (Dordrecht: Springer, 2007).

Rose, F. C., *Twentieth-Century Neurology: The British Contribution* (London: Imperial College Press, 2002).

Rudmose, B. R. N., 'Sir William Abbott Herdman (1858–1924)', *Oxford Dictionary of National Biography* (Oxford: Oxford University Press, 2004).

Rupke, N., *Richard Owen: Biology without Darwin* (Chicago, IL: University of Chicago Press, 2009).

Ruse, M., 'Woodger on Genetics, a Critical Evaluation', *Acta Biotheoretica*, 24 (1975), pp. 1–13.

—, *The Darwinian Revolution: Science Red in Tooth and Claw* (Chicago, IL: University of Chicago Press, 1999).

Russell, E. S., 'Évolution ou épigénèse', *Scientia*, 8 (1910), pp. 225–46.

—, 'Vitalism', *Scientia*, 9 (1911), pp. 329–45.

—, 'Review of *Sur la Transmissibilité des Caractères Acquis: Hypothèse d'une Centro-épigénèse*', *Scientia*, 11 (1912), p. 439.

—, 'Bateson, W. – Mendel's Principles of Heredity', *Scientia*, 15 (1914), pp. 274–8.

—, 'Vererbungslehre', *Scientia*, 15 (1914), pp. 278–81.

—, *Form and Function: A Contribution to the History of Morphology* (London: J. Murray, 1916).

—, *The Study of Living Things: Prolegomena to a Functional Biology* (London: Methuen, 1924).

—, *The Interpretation of Development and Heredity* (Oxford: Clarendon Press, 1930).

—, 'Fishery Researches: Its Contribution to Ecology', *Journal of Ecology*, 20:1 (1932), pp. 128–51.

—, *The Directiveness of Organic Activities* (Cambridge: Cambridge University Press, 1945).

Safranski, R., *Romantik. Eine deutsche Affäre* (Munich: Hanser, 2007), trans. M. Ritterson, 2008, *Goethe Institute, German Literature Online*, at http://www.litrix.de/mmo/priv/24106-web.pdf [accessed 9 August 2013].

Sander, K., 'Wilhelm Roux and the Rest: Developmental Theories 1885–1895', *Roux's Archives of Developmental Biology*, 200 (1991), pp. 293–9.

Sapp, J., *Beyond the Gene: Cytoplasmic Inheritance and the Struggle for Authority in Genetics* (Oxford: Oxford University Press, 1987).

—, *Genesis: The Evolution of Biology* (Oxford: Oxford University Press, 2003).

—, 'Just in Time: Gene Theory and the Biology of the Cell Biology', *Molecular Reproduction and Development*, 76 (2009), pp. 1–9.

Sarton, G., 'Eugenio Rignano', *ISIS*, 15:1 (1931), pp. 158–62.

Schadewaldt, H., 'Leuckart, Karl Georg Friedrich Rudolf', *Encyclopedia.com*, at http://www.encyclopedia.com/topic/Karl_Georg_Friedrich_Rudolf_Leuckart.aspx [accessed 13 August 2013].

Schelling, F. W. J., *Werke* (Leipzig: Band 1, 1907).

—, *Escritos sobre la filosofía de la naturaleza*, ed. A. Leyte (Madrid: Alianza Editorial, 1996).

Schwenk, K., D. K. Padilla, G. S. Bakken and R. J. Full, 'Grand Challenges in Organismal Biology', *Integrative and Comparative Biology*, 49:1 (2009), pp. 7–14.

Schmidt, K. P., 'Warder Allee 1885–1966', in *Biographical Memoirs: National Academy of Sciences* (Washington, DC: National Academies Press, 1957), at http://www.nasonline.org/publications/biographical-memoirs/memoir-pdfs/allee-warder.pdf [accessed 13 August 2013].

Schwarz-Weig, E., 'Science in a Constant Flow: Life and Work of Eduard Strasburger (1844–1912)', *German Botanical Society*, 2002, at http://www.deutsche-botanische-gesellschaft.de [accessed 13 August 2013].

Seamon, D., and A. Zajonc, *Goethe's Way of Science: A Phenomenology of Nature* (Albany, NY: SUNY Press, 1988).

Sedgwick, A., *Student's Text-Book of Zoology* (London: Macmillan, 1898).

Shor, E. N., 'How Scripps Institution Came to San Diego', *Journal of San Diego History*, 27:3 (1981), pp. 161–73.

Sloan, P. R., 'Kant and British Bioscience', in P. Huneman (ed.), *Understanding Purpose: Kant and the Philosophy of Biology* (Rochester, NY: University of Rochester Press, 2007), pp. 149–70.

Smocovitis, V. B., 'Unifying Biology: The Evolutionary Synthesis and Evolutionary Biology', *Journal of the History of Biology*, 25 (1992), pp. 1–65.

—, *Unifying Biology: The Evolutionary Synthesis and Evolutionary Biology* (Princeton, NJ: Princeton University Press, 1996).

Smuts, J. C., *Holism and Evolution* (New York: Macmillan, 1926).

Spemann, H. 'Experimentelle Forschungen zum Determinations und Individualitäts Problem', *Naturwissenschaften*, 7 (1919), pp. 581–91.

Stephens, L. D., 'Joseph Leconte's Evolutional Idealism: A Lamarckian View of Cultural History', *Journal of the History of Ideas*, 39:3 (1978), pp. 465–80.

Stevens, M. H., 'The Enigma of Meyer Lissner: Los Angeles's Progressive Boss', *Journal of the Gilded Age and Progressive Era*, 8:1 (2009), pp. 111–36.

Stonequist, E. V., 'Eugenio Rigano 1870–1930', *American Journal of Sociology*, 36:2 (1930), pp. 282–4.

Stork, O., 'Berthold Hatschek: A Landmark in the History of the Morphology of the Austrian Academy of Sciences', *Almanac for the year*, 99 (1946), p. 284.

Strasburger, E., *A Text-Book of Botany* (London: Macmillan, 1898).

Sturdy, S., 'Biology as Social Theory: John Scott Haldane and Physiological Regulation', *British Journal of the History of Science*, 21 (1988), pp. 315–40.

Sumner, F. B., 'William Emerson Ritter: Naturalist and Philosopher', *Science*, new ser., 99:2574 (1944), pp. 335–8.

Sunderland, M. E., 'Regeneration: Thomas Hunt Morgan's Window into Development', *Journal of the History of Biology*, 43:2 (2010), pp. 325–61.

Taquet, P., *Georges Cuvier, Naissance d'un Génie* (Paris: Odile Jacob, 2006).

Temkin, O., 'German Concepts of Ontogeny and History around 1800', *Bulletin of the History of Medicine*, 24:3 (1950), pp. 227–46.

Thieffry, D., 'Rationalizing Early Embryogenesis in the 1930s: Albert Dalcq on Gradients and Fields', *Journal of the History of Biology*, 34 (2001), pp. 149–81.

Thomson, J. A., *The Study of Animal Life* (London: J. Murray, 1901).

—, *Systems of Animate Nature: The Gifford Lectures Delivered in the University of St. Andrew in the Years 1915 and 1916*, 2 vols (London: Williams and Norgate, 1920).

—, and P. Geddes, 'A Biological Approach', in J. E. Hand (ed.), *Ideals and Science and Faith* (London: George Allen, 1904).

Thompson, D. W., 'The Regeneration of Lost Parts in Animals', *Mind*, 9:35 (1884), p. 419.

—, 'Review of J. S. Haldane's The New Physiology', *Mind*, 27 (1919), pp. 359–61.

—, *On Growth and Form* (Cambridge: Cambridge University Press, 1942).

—, *On Growth and Form*, ed. J. T. Bonner, rev. and abridged edn (Cambridge: Cambridge University Press, 1966).

Thompson, R. D., *D'Arcy Wentworth Thompson, The Scholar-Naturalist, 1860–1948* (Oxford: Oxford University Press, 1958).

Thone, F. E. A., and E. W. Bailey, 'William Emerson Ritter: Builder', *Scientific Monthly*, 24 (1927), pp. 256–62.

Tilliette, X., *Schelling: une philosophie en devenir* (Paris: Vrin, 1970).

Ungerer, E., *Die Teleologie Kants und ihre Bedeutung für die Logik der Biologie* (Berlin: Verlag von Gebrüder Borntraeger, 1922).

Van der Weele, C., *Images of Development: Environmental Causes in Ontogeny* (Albany, NY: SUNY Press, 1999).

Von Aesch, A. G., *Natural Science in German Romanticism* (New York: Columbia University Press, 1941).

Von Baer, K. E., 'Fragments Related to Philosophical Zoology: Selections from the Works of K. E. von Baer, trans. T. H. Huxley', in A. Henfrey and T. Huxley (eds), *Scientific Memoirs, Selected from the Transactions of Foreign Academies of Science* (London: Taylor & Francis, 1853), pp. 176–238.

Von Bertalanffy, L., *Modern Theories of Development: An Introduction to Theoretical Biology* (Oxford: Oxford University Press, 1933).

—, W. R. Ashby and G. M. Weinberg, *Trends in General System Theory* (Hoboken, NJ: John Wiley & Sons, 1972).

Von Molnar, G., 'Goethe's Reading of Kant's Critique of Esthetic Judgment', *Eighteenth Century Studies*, 15 (1982), pp. 402–20.

Wallace, A., *A Theory of the Evolution of Development* (Hoboken, NJ: John Wiley & Sons, 1988).

—, 'D'Arcy Thompson and the Theory of Transformation', *Nature Review Genetics*, 7:5 (2006), pp. 401–6.

Watson, D. M. S., 'James Peter Hill, 1873–1954', *Biographical Memoirs of Fellows of the Royal Society*, 1 (1955), pp. 101–17.

Weismann, A., *The Germ-Plasm: A Theory of Heredity* (New York: C. Scribner's Sons, 1988).

Wellek, R., *Immanuel Kant in England, 1793–1838* (Princeton, NJ: Princeton University Press, 1931).

Werskey, G., *The Visible College* (London: Penguin, 1978).

Whitehead, A. N., *Science and the Modern World* (New York: Free Press, 1997).

Whitfield, J., *In the Beat of a Heart: Life, Energy and the Unity of Nature* (Washington, DC: Joseph Henry Press, 2006).

Whitman, C. O., 'The Inadequacy of the Cellular Theory of Development', *Journal of Morphology*, 8:3 (1893), pp. 639–59.

Willier, B. H., 'Frank Rattray Lillie, 1870–1947', in *Biographical Memoirs: National Academy of Sciences* (Washington, DC: National Academies Press, 1957), at http://www.nasonline.org/publications/biographical-memoirs/memoir-pdfs/lillie-frank-r.pdf [accessed 13 August 2013].

Wilson, E. B., *The Cell in Development and Inheritance* (London: Macmillan, 1896).

Winchester, S., *The Man Who Loved China: The Fantastic Story of the Eccentric Scientist Who Unlocked the Mysteries of the Middle Kingdom* (New York: Harper Editions, 2008).

Winsor, P. M., *Reading the Shape of Nature: Comparative Zoology at the Agassiz Museum* (Chicago, IL: University of Chicago Press, 1991).

Wolff, M., 'Hegel's Organicist Theory of the State: On the Concept and Method of Hegel's Science of the State', in R. Pippin (ed.), *Hegel on Ethics and Politics* (Cambridge: Cambridge University Press, 2004), pp. 291–322.

Woodger, J. H., 'On the Relationship between the Formation of Yolk and the Mitochondria and Golgi Apparatus during Oogenesis', *Journal of the Royal Microscopical* Society, 40:2 (1920), pp. 129–56.

—, 'Notes on a Cestode Occurring in the Haemocoele of House-Flies in Mesopotamia', *Annals of Applied Biology*, 3 (1921), pp. 346–51.

—, 'On the Origin of the Golgi Apparatus on the Middle-Piece of the Ripe Sperm of Cavia', *Quarterly Journal of Microscopal Science*, 65 (1921), pp. 265–91.

—, *Elementary Morphology and Physiology for Medical Students: A Guide for the First Year and a Stepping-Stone to the Second* (Oxford: Oxford University Press, 1924).

—, 'Observations on the Origin of the Germ-Cells in the Fowl, Studied by Means of their Golgi Bodies', *Quarterly Journal of Microscopal Science*, 69 (1925), pp. 445–62.

—, *Biological Principles: A Critical Study* (London: Kegan Paul, Trench, Trübner and Co., 1929).

—, 'The Concept of Organism and the Relation between Embryology and Genetics Part I', *Quarterly Review of Biology*, 5:1 (1930), pp. 1–22.

—, 'The Concept of Organism and the Relation between Embryology and Genetics Part II', *Quarterly Review of Biology*, 5:4 (1931), pp. 438–62.

—, 'The Concept of Organism and the Relation between Embryology and Genetics Part III', *Quarterly Review of Biology*, 6:2 (1931), pp. 178–207.

—, *The Axiomatic Method in Biology* (Cambridge: Cambridge University Press, 1937).

Zammito, J. H., *The Genesis of Kant's Critique of Judgment* (Chicago, IL: University of Chicago Press, 1992).

# INDEX

For Product Safety Concerns and Information please contact our EU
representative GPSR@taylorandfrancis.com
Taylor & Francis Verlag GmbH, Kaufingerstraße 24, 80331 München, Germany

www.ingramcontent.com/pod-product-compliance
Ingram Content Group UK Ltd.
Pitfield, Milton Keynes, MK11 3LW, UK
UKHW021008180425
457613UK00019B/858

* 9 7 8 1 1 3 8 6 6 2 2 8 5 *